The Physics of
GOLF

Second Edition

Theodore P. Jorgensen

 Springer

Theodore P. Jorgensen
Department of Physics and Astronomy
University of Nebraska, Lincoln
Lincoln, NE 68506
USA

Cover photo: Photonica/The Picture Book, copyright 1999.

Library of Congress Cataloging-in-Publication Data
Jorgensen, Theodore P. (Theodore Prey)
   The physics of golf / Theodore P. Jorgensen. -- 2nd ed.
       p. cm.
   Includes bibliographical references and index.
   ISBN 0-387-98691-X (alk. paper)
   1. Physics. 2. Golf. 3. Force and energy. I. Title.
QC26.J67 1999
796.352–dc21   98-32112

Printed on acid-free paper.

Production managed by Terry Kornak; manufacturing supervised by Thomas King.
Typeset by American Institute of Physics, Woodbury, NY.
Printed and bound by R. R. Donnelley and Sons, Harrisonburg, VA.
Printed in the United States of America.

9 8 7 6 5 4 3 2

ISBN 0-387-98691-X Springer-Verlag New York Berlin Heidelberg   SPIN 10753574

PREFACE TO THE SECOND EDITION

When I was introduced to the game of golf, I assumed that I would soon be hitting the little white ball about as well as anyone. Of course, I could not. I then went to the library to discover what the experts had to say. A careful reading of many books offered minimal help. These authors were not discussing fundamentals that had meaning to a physicist. At this point, I realized that my scientific curiosity about golf would be satisfied only by seeking the answers in my own research.

The need for a study of the kind I was to undertake had been recognized for quite some time. Bobby Jones, a well-known golfer of a former day, wrote, "There are a number of players who devote enough thought, time, and practice to the game to make them reasonably good golfers if they might start out with an accurate conception of what they want to do; but in many instances there is a confusion of ideas making intelligent progress impossible." Therefore, part of my intent was to show that the science of physics is able to provide some relief from this "confusion of ideas."

My study of the physics of golf began late one afternoon in my office when I posed this question: "Would an investigation into the physics of the swing of a golf club help improve my game?" My work-book entry, dated May 9, 1968, shows that I intended to use the differential equations of motion of a fairly simple double pendulum model of the swing of a golf club "to see whether any insight might be had into the nature of the optimum golf swing."

The swing of a golf club is indeed an example of what a physicist calls a double pendulum; the arms of the golfer act as one pendulum connected to the club, which acts as another pendulum. Methods of obtaining the equations of motion for a double pendulum had been worked out long before. However, no one could apply this theory to the swing of a golf club because calculations could not be made quickly enough to be practical. A computer capable of rapidly solving these equations became available to me only shortly after I had become interested in applying physics to problems in golf.

I wrote a paper published in the *American Journal of Physics* as a result of my first calculation of the double pendulum problem. The American Institute of Physics used this paper for public relations purposes because it was an interesting example of the application of physical principles in the field of sports. *The National Observer* featured my work in a front-page story. I received many requests for reprints, and still receive letters requesting answers to golf science questions.

At the time of my first paper there was no firm experimental work available with which to compare my calculations. The few stroboscopic photographs in the literature were taken with the camera off the swing axis and showed blurred pictures from which I could not obtain precise measurements.

Fortunately, at about this time the 3M Company came up with their highly superior reflecting tape, which allowed me to produce better stroboscopic photographs needed for my work. Older photographs were taken with an intense flashing light source, and the finished picture was simply many superposed photographs of the golfer. By using the reflecting tape and a much weaker flashing light source, and putting the tape only on certain locations of the golfer and the club, positions of these tapes would stand out clearly in the photograph. The photographs were taken from a position on the axis of the swing about 25 feet from the golfer. These photographs allowed me to make precise comparisons between theory (calculations) and experiment (photographs).

As I began to apply my understanding of physics and mathematics to the fundamentals of golf, I watched things being done by golfers, amateurs and professionals alike, that showed a lack of scientific understanding. I shared my ideas with friends, but the only way I had of reaching a large number of golfers was to write about my findings. This book is the product of my twenty-five years of research.

You will find this a book about golf unlike any other. It is not a "how to" book, nor a compilation of golf's greatest stories, but an insightful attempt to bring some fundamental understanding to the game of golf. My research has allowed me to know some of what I should be trying to do to improve my game. I hope that this book may bring some understanding to others, too.

Shortly after the publication of the first edition of this book, I began to receive feedback from readers, which ranged from enthusiastic explanations of what the book meant to them to expressions of disappointment from some who were not acquainted with enough of the language of elementary physics to benefit as much as they had expected. I had not assumed that so many of the golfing public were handicapped by not being acquainted with certain concepts that are necessary to a discussion of golf from this different point of view. In this second edition I have attempted a nonmathematical section explaining the elementary concepts of Newtonian dynamics basic to the reporting of this research. This material cannot be hurriedly read with much profit, but if it is studied slowly with thought, while it will not give a reader a working knowledge of the subject, it should facilitate an understanding of the Physics of Golf.

After the introductory chapter, Chapter 2 describes the fundamental approach to my research: how I find from a photograph what a professional golfer does when he swings a driver and how I establish a mathematical model using Newtonian dynamics that allows me to make calculations that give me the same clubhead speed curves as I obtained from the photograph. With the same mathematical expressions, I find that I am then able to calculate all sorts of things concerning the golf swing.

Chapter 3 explains how the golf swing compares to the action of a bullwhip, how the large muscles of the lower body, although they are not connected to the arms, are used to enhance the speed of the golf club, how the pull of the shoulder on the arm works in the golf swing, and how we find that the golfer does not swing about a quiet center in the downswing.

In Chapter 4, the calculated swing is examined by varying parameters in the equations. The method of varying parameters answers many questions about the swing such as the deleterious effect of an uncocking process by the wrists, the reason for using a shortened backswing, and the effect of the initial wrist cock at the top of the backswing on the clubhead speed at impact.

In Chapters 5 and 6, the concepts of energy and work are used in studying the golf swing. An examination of the general problem of producing a powerful stroke with precision shows the importance of the pull of the left shoulder on the left arm.

I had long considered whether I should include in this book a section on how a golfer might use the results of my research in developing his own golf stroke. At first I doubted that I, a man of many years who had not mastered the game of golf to his satisfaction, should attempt to coach golfers through the written word in this matter. However, several years into my research, I had some experiences which prompted me to write Chapter 7. I shall describe one of these which stands out in my memory.

I observed a young man hitting balls at a local driving range with what I learned later was a two iron. His swing was in a groove; his balls all sliced out over the south pasture. I watched him to see what his troubles might be. One thing was quite obvious; he ended his stroke with his weight on his right leg. After watching him for a while and thinking I might be of some help, I went down to him and asked him if he minded if I coached him a bit. He said he would appreciate it very much if I could help him.

I told him, as best I could, what he should be doing, but as a result of his habits, he was completely incapable of following my directions. At last I asked him if I might have his club in my hands for a moment to show him what I had in mind. I did not hit a ball for him. Rather I showed him in an exaggerated fashion something of what I learned from my five experiments described in Chapter 3. I tried to have him pull on his left arm by his left shoulder in the direction he wanted the ball to go. You cannot do this without shifting your weight. It was difficult for him to do this because such a motion was completely different from what he had been doing. I told him he was to pull so hard on the upper end of his left arm at the beginning of the down-swing that he could feel the force he was exerting. With his attention focused on feeling this force and forgetting about swinging the club, he finally was doing what I was trying to tell him to do.

After several swings to fix the idea in his mind, he finally placed a ball on the ground, and with his new swing, hit the ball far and straight without even a hint of a slice. He hit a second ball with the same result. He hit several balls that would make any golfer proud. Finding that his slice had left no trace, he turned to me with the question, "Is that all there is to it?" I did not spoil the moment for him by answering that indeed there was much more to it. I did tell him to practice the new stroke until it was habitual with him and I thought he would feel much better about his game.

Seeing the almost instantaneous effect of my coaching on this person's golf stroke gave me the needed encouragement to start on what became the chapter in question. I used ideas that came from my studies combined with ideas I had acquired from my reading in golf.

Chapters 8 and 9 explain why a golf ball in the air behaves as it does and how it gets its spin, which causes the lift on the ball and the forces responsible for slices and hooks. Chapter 10 tells a little detective story of how Harry Vardon, a historical figure of golf, swung his clubs with such exceptional results. Chapters 11 and 12 deal with the matching of clubs and the question of flexible shafts. And finally, Chapter 13 looks at the question of whether the handicap system does what it is purported to do. Chapter 14 is a short chapter on short putts. The book ends with a Technical Appendix.

It is my hope that this book will have some appeal for everyone. For the general reader I try to give a glimpse of how a study using the concepts of physics may produce a new and clearer understanding of the fundamentals of a sport. The reader who has just taken up the game or intends to do so, without even having an interest in the technical aspects of the study, should find that this book will help him recognize what is fundamental in the golf swing. If the reader has a friend who can critically observe his swings, he should find what he learns from this book to be a real help in becoming more satisfied with his game. And the advanced golfer who feels he has come about as far as possible may after careful study reach a deeper understanding of the technical aspects of the golf stroke, and, with help, bring his game to a still higher level. He will also need help because no golfer can see his own swing.

The select group of readers of this book who would like to do research in areas that have not been treated here will find the Technical Appendix along with a computer a great entry point into this fascinating area of research.

While the playing of golf does not depend on the sex of the golfer, I have used the masculine pronoun throughout this book to avoid the stylistic infelicities of he/she.

I would like to thank the 3M Company for graciously supplying me with some samples of their reflecting tape. I wish to express my appreciation to the two professional golfers Bob Schuchart and Jerry Fisher for swinging clubs for the stroboscopic photographs that made this work possible. I wish to thank Donal Burns and Tom Braid for their suggestions, encouragement, and help with the manuscript.

CONTENTS

Technical Appendix

CHAPTER 1

The Secret of Golf Is in the Swing

Golf, with its intense frustrations, bitter disappointments, sturdy enjoyments, and, yes, extreme ecstasies, would not be the game it is without the continual dream to do better the next time around. Golfers dream about herculean 300-yard drives, perfectly lofted iron shots from the rough, masterful wedge shots from sand traps, and dropping 40-foot putts. Spurred on by the exaggerated claims from equipment manufacturers, they become suspicious that there must be serious faults with their equipment that prevent the realization of these dreams. Visit any pro shop or cut-rate golf store and feast your eyes on the latest golfing wares. How can a golfer resist?

The equipment of the present day, both balls and clubs, has evolved through a long process. The first golf balls were called featheries. They were made with a horsehide cover packed with wet goose feathers [1], which when they dried became very hard. These balls could not be used in wet weather. Until the middle of the 19th century, the game of golf remained essentially static, but then the revolutionary gutta-percha ball, or "guttie," was invented [1]. This new ball was molded from the warmed, dried gum of the sapodilla tree. The expense of the featheries had kept the game of golf from being played extensively. The guttie ball could be made very cheaply, and golf became very popular [2]. The industry associated with the manufacture of featheries declined, and the business of making the new balls boomed. It was soon discovered that the new balls flew better as they became roughened in play. By 1900 the balls were manufactured with a surface covered by "brambles," giving them the appearance of large, white blackberries. The guttie ball probably could not be driven as far as a good feathery ball, but considering the vast difference in price, it became the ball of choice for most golfers. In a decade or so, featheries were found only in the possession of collectors.

A remarkable event occurred in 1900 that possibly can be heralded as the first professional sports endorsement. The Spalding Company paid England's Harry Vardon a considerable sum of money to come to the United States to demonstrate what he could do in winning tournaments using the latest gutta-percha ball. He won the 1900 U.S. Open.

The almost immediate acceptance of the new rubber-wound Haskell

1

ball, invented in 1898, caused the demise of the guttie ball. This invention produced another revolution in golf, because this ball could be hit farther. At the time of its introduction, there was talk that it should be banned from tournament play; its acceptance would spoil the game, and golf courses would have to be lengthened. However, distance off the tee won out, and the rubber-wound ball came into universal use.

It was not until 1968 that a new, two-piece ball came on the market. The two-piece ball had a plastic rubber core covered with a new material, Surlyn, that was more durable than the balata cover of the later Haskell balls. The two-piece ball immediately found a large niche; its harder surface gave less spin to the ball, and with less spin, there was less drag, and the ball could be hit farther than the rubber-wound ball. The new cover material was practically cut-proof. With less spin, the ball would produce smaller hooks and slices. The golf ball continues to be developed with the new understanding of aerodynamics of dimpling and of the effects of the ball's structure on spin. However, the rubber-wound ball with a balata cover is almost universally used in professional tournament play.

While balls were being developed, clubs were also changing. First, metal heads replaced the wooden heads when guttie balls came into use (iron clubs were too damaging to be used with the featheries). Next, hickory shafts became universal. There were some experiments with steel shafts, but such shafts were not allowed by the United States Golf Association (USGA) until 1926. Bobby Jones continued to use hickory shafts until the 1930s.

Some golfers began to carry close to 30 clubs: a number of irons, numerous woods, a left-handed club, special wedges, and others. In 1937, before things got completely out of hand, the USGA set standards for golf balls and clubs. The number of clubs a golfer could carry for competition play was limited to 14. These standards were general enough to allow such clubs as the sand wedge and the pendulum putter.

As more people became interested in golf, more courses were built, both private and public, and this process continues. The economic impact of golf is reflected in the recent estimate of $15 billion spent annually on this sport.

There was some stumbling along the way in the development of golf equipment. Some changes in equipment quickly became widely accepted, while others quickly faded. No balls of today have brambles. One-piece balls are now a curiosity. Steel shafts are here to stay, while some shafts of other materials, although introduced with considerable publicity, have disappeared.

One well-known golfer was paid a handsome sum by a manufacturer to point out the virtues of clubs made with fiberglas shafts. The golfer used these clubs to win a tournament. A major selling feature of these shafts was that they were purported to damp out the flutter of the clubhead produced by inertial forces at the start of the downswing, a supposedly serious problem that probably had no existence except in the imagination of the manufacturer. Clubs with fiberglas and aluminum shafts are no longer found in shops. One relatively new development in clubheads, perimeter weighting, which increases the moment of inertia of the clubhead about an axis parallel to the shaft, is based on an idea having a firm theoretical basis and is likely here to stay.

The drive of a golfer to improve his game makes him vulnerable to the glamour used in marketing the new equipment. Fashion probably also plays some part. When metal woods appeared, the urge to buy these clubs must have been almost irresistible. While it is true that these clubs are less destructible than wooden clubs and they have perimeter weighting, the stampede to these clubs likely came more as a result of an intense marketing campaign than as a result of the metal woods' superiority.

Certainly, the improvement of golfing equipment is real. Golf balls have had their aerodynamic properties improved by their dimpled surfaces. The hickory shafts are long gone. Part of the lure of present-day equipment may be aesthetic, but in any case, it is doubtful that anyone would like to return to the equipment of earlier years.

Some studies have been made to find an answer to the question of whether the equipment a golfer uses has a significant influence on his game. Such an influence should show up in the statistics of the Professional Golfers' Association (PGA). The PGA has collected statistics for some years [3]. In the period from 1964 to 1989, in which significant changes in equipment occurred, the median length of drives increased by six yards. Such an increase could as well be attributed to the general increase in professional golfers' distance off the tee, resulting from a general increase in skill from improved diet, fitness training, and more competition. In comparison to other sports, this increase is minimal. Another measure of improvement is in the scores of these professional golfers. The median score of golfers completing the four rounds in the U.S. Open Championships has been dropping at a rate of about two strokes per ten years. For ordinary golfers, the average USGA handicap from 1979 to 1990 has remained a steady 17.

Thus the reason for this book may be briefly stated: "It is a poor workman who curses his tools." Any golfer who wishes to improve his game must work on himself. He is the one who makes his clubs go through their motions. The secret of golf is in the swing. As one knowledgeable golfer put it, "The effect of equipment is found to be in small percentages here and small percentages there, but these never add up to very much."

CHAPTER 2

A First Look at the Golf Stroke

The Swing of the Club

A golf stroke by an accomplished golfer is usually a thing of beauty. The motion of the golfer may be executed with such precision that the ball flies just as the golfer expected, in the right direction and for the correct distance, and rolls to stop near the point intended. However, instead of considering the aesthetics of a golf stroke, our intention is to look at it in another way, by considering it as a dynamic process using the physical principles of mechanics based on Newton's laws of motion. The stroke may be considered as a sequence of three separate events: the swing of the club, the impact of the clubhead with the ball, and the flight of the ball toward the target. We shall start our application of physics to golf by looking at the first of these, the swing of the club.

If you watch your fellow golfers, or even the professionals on television, you soon realize the infinite variety in the swing of a golf club. Some swings are a delight to watch and are probably a great satisfaction to the golfer. However, some swings are so grotesque that the exasperation expressed comes as no surprise. It would be useless to try to analyze all possible swings. Rather, our attention will be directed toward a particular swing by a professional golfer, and we shall try to match this swing with one calculated according to a simple model.

Scientific Concepts are Necessary for Understanding

In the material of this chapter and several others we shall of necessity be using fundamental concepts of dynamics, the theory of moving bodies, which is studied in elementary physics. It was not long ago that these concepts, without which motion cannot be understood, did not exist. They developed slowly over the past five hundred years. The names of two men of genius, Galileo Galilei and Isaac Newton, stand out. Galileo studied motion, and Newton is famous for his laws of motion. Galileo admitted that he did not understand how a hammer produces a large force. Newton's laws of motion provided an under-standing of linear and rotational motion. Remember that we are still trying to understand the swing of a golf club.

The concepts we shall use are speed, linear velocity, linear accelera-tion, angular velocity, angular acceleration, momentum, angular momentum, mass, moment of inertia, force, torque, centripetal force, and centrifugal torque. Newton's laws of linear motion and their exten-sion to rotational motion are basic to the development of our under-standing of the swing of a golf club. If the reader is familiar with these laws and most of these concepts, he should not be troubled in reading these chapters. If he is not comfortable with his knowledge in this area, he is advised to read and study the discussion of Newtonian dynamics in Section 9 of the Technical Appendix. In any case, this material is provided for the convenience of the reader.

The Stroboscopic Photograph

The details of a golfer's swing were obtained from a stroboscopic photograph. We needed a permanent record showing details of the swing as it varied with time. One method of stroboscopic photography is exemplified by the picture of Bobby Jones swinging a driver, as shown by Harold E. Edgerton in *Flash* [4]. Such photographs were essentially many photographs taken with a bright flashing light super-posed on one another. These photographs show a general blur over most of the picture and are taken with the camera off the axis of the swing. It is difficult to obtain reliable data from such photographs. Therefore, it was necessary to devise a method for taking stroboscopic photographs of a golfer that allow a better determination of the data for the actual swing of a golf club. The details of this method are described in the Technical Appendix, Section 7.

The original stroboscopic photograph was a 35 mm slide; an enlarge-ment of this slide is shown in Fig. 2.1. The spots in the photograph were produced by light from a flashing source reflected from highly reflecting tape placed at various locations on the golfer and on the club. To obtain data for the swing, the slide was projected on a large sheet of paper, and the relevant spots on the image were marked with a pen. The distances between spots were measured. The slide has calibration spots on it separated by a known distance at the position of the golfer. With a known flash rate for the stroboscopic light source, it is a fairly simple matter to find the speeds at which the tapes were moving.

During the downswing, the club turns about its shaft, and since the clubhead is offset, a bit of tape on the clubhead would not appear to be on the shaft. For this reason, the position of the clubhead was not determined directly. Rather, a bit of tape was put on the shaft directly

FIGURE 2.1. A professional golfer swung his driver for this stroboscopic photograph. Each dot on the photograph shows the position of a piece of highly reflecting tape on the golfer or on the golf club. The four dots at the bottom of the photograph are for calibration purposes. The curve of rectangular dots farthest to the left in the photograph are from a tape on the head of the club, and these dots were produced during the backswing of the golfer. The corresponding three curves of dots were produced by tapes on the shaft of the club, one tape next to the clubhead, one on the club at the grip, and one on the club halfway between these two. The curves of these dots lead into the top of the backswing, where they move into the corresponding curves of dots in the downswing. The stroboscopic flashes came at equal time intervals. Consequently, the distances between successive dots on a given curve are proportional to the average speed of the tapes for these time intervals. The curves to the right of the photograph were made when the golfer was in his follow-through. The triangular figure at the center of the photograph was made by light reflected from a tape on the head of the golfer. The shape of this curve is as unique for a particular golfer as his signature.

adjacent to the clubhead. We shall call this bit of tape Tape A. As may be seen from Fig. 2.1, the position of Tape A could be determined with suitable precision.

All the experimental data for the speed of Tape A are given in Table I and are also shown by dots in Fig. 2.2, where they are plotted on a speed–time curve. These data were plotted on a large scale, and a line was drawn that gave a good fit to the data. Speeds of Tape A were read from this curve at intervals of one-hundredth of a second. These speeds

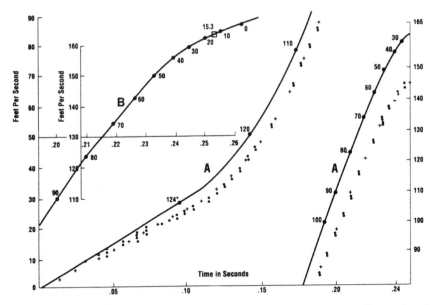

FIGURE 2.2. (A) The curves show the experimentally determined speeds of Tape A, the tape on the shaft adjacent to the clubhead, and the speeds of Tape A and the clubhead determined by the computer calculation of the Standard Swing. The dots are the plots of the averages of the four independent determinations of the speed of Tape A. The crosses (+) are plots of the computer calculations of speeds of Tape A. The curve marked A is a plot of the computer calculations of speeds of the clubhead. The circled point of this curve are the speed–time values for the particular wrist cock angles of these points. The number near each point is its wrist cock angle in degrees. (B) The curve shows the clubhead speeds determined for the Standard Swing plotted to a different time scale. This curve shows the curvature at the high clubhead speed section better than the scale used in (A). The circled points on the curve are the speed–time values for particular wrist cock angles for these points. The number near each point is its wrist cock angle in degrees. The point within the square indicates the speed–time where the clubhead hits the ball.

are listed in Table II. Tables I and II are found in the Technical Appendix, Section 8.

There are inherent errors in this method of determining the speed of Tape A, as it depends on time into the downswing. The images on the slide showing the positions of Tape A at the beginning of the downswing are so close together that no useful data can be obtained for this part of the downswing. This means that the time at which the downswing begins can be determined only by an uncertain extrapolation of the speed–time curve. Also, at the end of the downswing the images on the slide are so far apart that any details of the slowing of the golfer's hands are not available. As in any physical measurement, there are observational errors to be considered, too.

Choice of Model

Usually, little progress can be made in understanding a physical event, such as the swing of a golf club, if the event is initially considered in all its complexity. The investigator must choose a simplified model to simulate the event. The model must be simple enough to be manageable but not so simple that it does not contribute to understanding the original event. The model chosen is a rather simple one for three reasons: The mathematical development remains fairly simple; most golfers use clubs with shafts of little flexibility; and it is the present style to swing a club with a straight left arm. However, as we shall see, this model does allow us to develop considerable understanding of the swing of a golf club.

The swings of golf clubs by competent golfers have certain common characteristics. The arms of the golfer swing about an axis that moves during the downswing, and the club swings about a moving axis near the wrists of the golfer. The essential elements of this mechanical model will then be two rigid rods, A (arm) and C (club), joined at a hinge representing the wrists of the golfer with the upper end of rod A constrained to rotate about an axis that moves horizontally toward the target during the downswing. The axis at the upper end of rod A is near the point halfway between the golfer's shoulders. Rod A may then be considered along a line from this axis to the golfer's wrists. A diagram of such a model is shown in Fig. 2.3. During the downswing, the axis at the upper end of rod A will be assumed to have an acceleration a, first positive toward the target and later negative as it slows and stops.

This model is more complicated from that considered by other investigators and by me in my earlier work. The simpler model neglects the possible accelerated shift of the center of rotation of the system and assumes that during the downswing the arms and therefore the hands move on a circular path about a fixed pivot. Later, in Chapter 3, we shall see that there are physical reasons why a knowledgeable golfer likely uses the shift toward the target during the downswing.

It will be assumed that the downswing starts at time $t = 0$ when rod A has been swung back from the vertical through a clockwise angle, the backswing angle, which we shall call gamma (γ). During the backswing and during the downswing the wrists are cocked, with rod C swung from the line that is an extension of rod A through a clockwise angle, which will be called the wrist-cock angle beta (β). We shall be interested in two particular values of beta. The value of beta at the start of

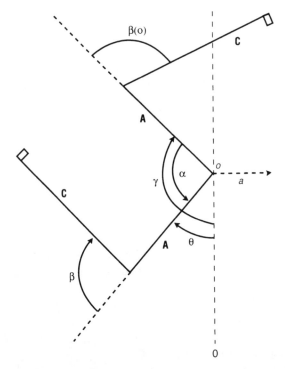

FIGURE 2.3. This is a diagram of the model chosen to represent a golfer's swing of a golf club. Rod A represents the arms of the golfer, having the equivalent mechanical properties of his two arms, and rod C represents the golf club, having the mechanical properties of the club used in the swing. The joint between rod A and rod C is flexible. The system rotates about a center at O, which has a horizontal acceleration a. At the beginning of the downswing, rod A is at an angle gamma (γ) with the downward direction, and the wrist-cock angle beta (β) has the value $\beta(0)$. At any time in the downswing, rod A is at an angle theta (θ) from the downward direction having moved through the downswing angle alpha (α) from its initial value $\alpha = 0$, and the wrist-cock angle has the value β.

the downswing is indicated by $\beta(0)$, beta at time $t = 0$. The value of beta when the clubhead is in contact with the ball will be indicated as $\beta(i)$, beta at impact. We shall call the angle that rod A makes from the start of the downswing at any time alpha (α). The time rate of change of this angle, the angular velocity, we shall call "alpha dot" $(\dot{\alpha})$, and the time rate of change of $\dot{\alpha}$, the angular acceleration of alpha, we shall call "alpha double dot" $(\ddot{\alpha})$. This same notation will be used to designate the angular velocity and the angular acceleration of beta.

As the downswing develops, through the action of a torque on the golfer's arms by his body, there are additional torques acting on the arms and the club. The effect of these torques on the motion of the arms and club can be analyzed according to the laws of dynamics, and exact mathematical equations may be written describing certain aspects of this motion. Even though it will seem in the following discussion that the equations are concerned with the motion of the arms of the golfer and the club, I must emphasize that the mathematical equations are descriptive of the motion of the model and not the motion of an individual golfer. But I hope that the results of the study of the model will be useful in understanding what goes on in the actual swing of a club by a golfer.

Torques on the Club

In applying the fundamental principles of dynamics to the downswing of a golf club using this model, we need two mathematical equations. They tell what is happening to the club and to the golfer's arms at any time during the downswing. The club may have five different torques acting on it. The arms of the golfer exert a torque on the club that gives it an accelerated rotational motion about the axis of the swing at O. The wrists, under the control of the golfer, may exert a torque on the club that may accelerate the uncocking motion of the wrists. The centrifugal torque experienced by the club, when looked at in the rotating system, also accelerates the uncocking process. A fourth torque comes from the force of gravity acting on the club. A fifth torque results from the acceleration by the golfer of the center of rotation in his shift toward the target.

Torques on the Arms

The golfer's arms may also have five different torques acting on them. At the beginning of the downswing, the largest of these is the torque by the golfer's body on his arms about their axis of rotation. If the golfer exerts a torque on the club by his wrists, which produces an acceleration, the club exerts an equal and opposite torque back on his wrists and thus on his arms. The rapid rotation of the club toward the end of the downswing produces a torque back on the golfer's arms that acts to slow their motion. The pull of gravity produces a torque on the arms. The fifth torque on the arms results from the acceleration by the golfer of the center of rotation in his shift toward the target. The fifth torques on the club and on the arms will be discussed more fully in Chapters 3 and 6.

The Mathematical Equations

The mathematical equations that describe these torques look somewhat complicated (see the Technical Appendix, Section 4), but the idea used in writing these equations is quite simple. One equation says that the sum of the five torques acting on the club is proportional to the angular acceleration of the club. The other says that the sum of the five torques acting on the arms is proportional to the angular acceleration of the arms. These are mathematical expressions of Newton's second law of motion applied to rotational motion. The constants in the equations depend on the dynamic parameters of the club and of the arms of the golfer. These constants depend on the mass and the mass distribution of each of them.

These mathematical expressions describe any swing of any club by any golfer according to our model. In order to have them describe a particular swing, we need to specify the dynamic properties of the club being swung and the dynamic properties of the golfer's arms, as well as can be determined without dismantling the golfer. We need to specify the manner in which the golfer uses his wrists, the manner in which he moves his center of rotation toward the target, and finally, the manner in which the golfer is going to exert the torque on his arms.

The Standard Swing

The quality and usefulness of our model will be determined by how closely our calculations match the experimentally determined relation between the speed of Tape A and the time into the downswing for a properly swung club. As we shall see, we are able to do this with a surprising degree of precision.

A computer was programmed to solve the two simultaneous differential equations produced by the model. The parameters of the golf club were known, and the parameters of the golfer's arms were first estimated from measurements on the arms. The other descriptive factors of the swing such as the torque exerted by the golfer on the system, the wrist-cock angle $\beta(0)$ at the beginning of the downswing, and the torque exerted by the golfer's wrists on the club during the swing were ultimately chosen to give the best match obtainable between the calculated swing and the experimental swing. The relation between the calculated speed of Tape A and the time into the downswing is shown in Fig. 2.2 by crosses (+) and in Table II, located in the Technical Appendix, Section 8. The speed of the clubhead as it depends on the time into the downswing may be determined by the same calculation giving the speed of Tape A. We shall call this particular calculated swing the standard swing.

Choice of Parameters

Except for the parameters of the golf club, the parameters chosen for the calculation of the standard swing cannot be considered unique. The effect of a slight variation in one parameter on the calculation of the speed–time curve for Tape A may be approximately countered by a slight variation in one or more of the other parameters. The parameters chosen for the standard swing appear to be reasonable. As may be seen in Table II, the values for the calculated speed of Tape A and the smoothed measured speed of Tape A are practically the same except for a slight difference in the first tenth of a second into the downswing. This difference was traced to the effect of gravity and the horizontal acceleration, which shows up in the calculations but is unobservable in the crowded images in the photograph at the beginning of the downswing.

At the top of the backswing the photograph shows the shaft of the club to be about 20 deg below the horizontal; this means that the sum $\beta(0) + \gamma$ is 290 deg. In the process of finding the parameters for the standard swing, this sum must remain unchanged.

It is extremely difficult to determine the precise torque applied to the system by the golfer during the downswing. This torque, TS (equivalent torque on the system) was determined by finding through trial the torque for which the calculated and smoothed speeds of Tape A agreed at the first one-tenth of a second into the downswing. The value of TS determined this way was about 57 lbs·ft. An estimate of the torque a golfer is able to apply was made by having him exert a force on a calibrated spring balance in a horizontal direction with his two hands. Without undue exertion he was able to press with a force of about 23 lbs, which with a radius of 2.2 ft, the length of the golfer's arms, gives a torque of about 51 lbs·ft. The professional golfer who swung the club for the stroboscopic photograph probably would be able to do better than the 23 lb press. I consider the 57 lbs·ft torque very reasonable from this crude experiment. The value TS for the remainder of the swing could be adjusted in the computer program as needed to fit the experimental results. It was surprising that the torque of 57 lbs·ft sufficed for the remainder of the downswing. This fact tells us that the golfer who swung the club used an essentially constant torque throughout the swing. We know of no reason why this should be true.

Computing Other Quantities

Besides the clubhead speed of the standard swing shown in Fig. 2.2, the computer was programmed to give other quantities characteristic of the standard swing as they depend on time into the downswing. These quantities are the wrist-cock angle with its rate of change and its acceleration and the angle through which the arms have moved with its rate of change and its acceleration. The numerical values of these other quantities will not be presented here but will be plotted in Figs. 2.4 and 2.5.

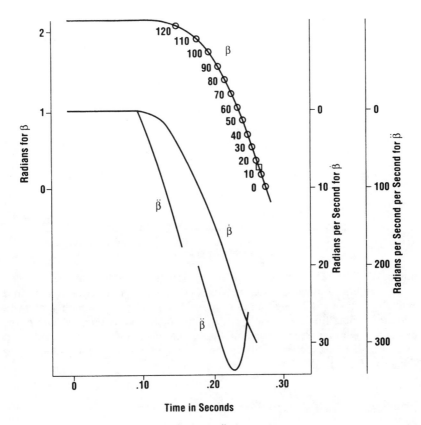

FIGURE 2.4. The values of β, $\dot{\beta}$, and $\ddot{\beta}$, the wrist-cock angle, its angular velocity, and its angular acceleration, during the standard swing. The curve shows a constant value of about 124 deg until the wrists start to uncock about one-tenth of a second into the downswing and then gradually decreases to zero. The circled dots indicate the values of β in multiples of 10 deg as the wrists gradually uncock. The dot in the square indicates the wrist-cock angle when the ball is hit. The curve for $\dot{\beta}$, the angular velocity of the wrist-cock angle, has the value zero for the first one-tenth of a second into the downswing and then has an ever increasing negative value. The curve for $\ddot{\beta}$, the angular acceleration of the wrist-cock angle, has the value zero for the first one-tenth of a second into the downswing and then shows the gradually increasing negative value of $\ddot{\beta}$. The discontinuity in the $\ddot{\beta}$ curve is related to the discontinuity in the acceleration of the golfer's shift a.

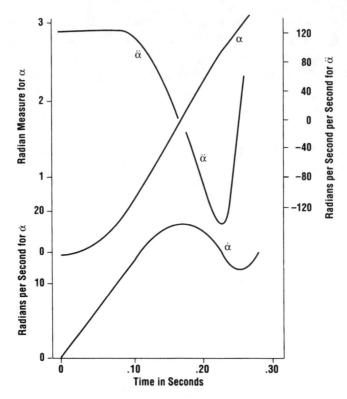

FIGURE 2.5. The values of α, $\dot{\alpha}$, and $\ddot{\alpha}$, the downswing angle, its velocity, and its acceleration, during the downswing. The curve for α starts at zero and gradually increases throughout the swing. The curve for $\dot{\alpha}$ starts at zero, increases almost linearly toward a maximum, and then decreases. The curve for $\ddot{\alpha}$ starts with an almost constant value, then decreases to a minimum, and then increases. The discontinuity in the $\ddot{\alpha}$ curve is related to the discontinuity of the golfer's shift a.

The Wrist-Cock Angle

First let us look at the curve for the wrist-cock angle β in Fig. 2.4. The calculation shows that this curve starts out with $\beta(0)$ equal to 2.167 *rad* (124.2 deg) and maintains this value for about one-tenth of a second. During this time the golfer exerts a gradually decreasing torque by his wrists, which maintains this constant wrist-cock angle. After one-tenth of a second into the downswing, the golfer relaxes his wrists so that he no longer exerts a torque by his wrists, and the centrifugal torque in the rotating system produces a gradually decreasing wrist-cock angle β, first at a slow rate and later at a very rapid rate. The small circles on this curve and the numbers, multiples of ten, indicate the positions on the curve at which the wrist-cock angles are 120 and 110 deg, and so on down to zero. The square on the curve indicates the value of $\beta(i)$, about 15.3 deg, when the clubhead is in contact with the ball. The significance of this particular value of $\beta(i)$ will be discussed later.

The Clubhead Speed

We have used the agreement between the calculated and experimental speed of Tape A to arrive at equations we may use to calculate other quantities of interest in the downswing. We are not particularly interested in these other quantities. The quantity that tells us the effect the swing will have at impact with the ball is, of course, the clubhead speed. This clubhead speed as calculated by the computer is plotted in Fig. 2.2, curve A. This line is seen to start off with an almost straight portion, as it should if the golfer exerts a constant torque on the system and the moment of inertia of the system remains constant. This portion lasts for about one-tenth of a second. Then, as the wrists uncock, the line curves upward into another nearly straight-line portion, and finally the slope of the curve falls off a bit toward the end of the downswing. As in the plot of the wrist-cock angle, small circles with numbers indicate the clubhead speeds for particular wrist-cock angles. Notice that the differences in clubhead speeds between the points that differ by 10 deg in the wrist cock angle become smaller and smaller as these angles tend toward zero. An enlarged portion of this curve is shown in the upper left of Fig. 2.2.

The clubhead speeds for the wrist-cock angles of 10, 20, and 30 deg are 1.0%, 2.2%, and 3.8%, respectively, less than the clubhead speed for the wrist-cock angle of 0 deg. In Chapter 4, the significance of these angles will be shown to be of special interest.

The Arm Motion

The golfer's arms and therefore his hands slow up just before the ball is hit. This curious motion has been known for quite some time but is not generally understood. We shall first consider the motion of the arms. The results of the calculations concerning this motion are shown in Fig. 2.5, where the data for α, $\dot{\alpha}$, and $\ddot{\alpha}$ are plotted as functions of the time into the downswing.

For the first one-tenth of a second, the graphs of the position angle of the arms α, its time rate of change $\dot{\alpha}$, and its acceleration $\ddot{\alpha}$ have the characteristics expected of a rigid body moving under the action of a constant torque. The value of $\ddot{\alpha}$ remains essentially constant during this time, the curve for $\dot{\alpha}$ is essentially a straight line, and the curve for α has essentially the shape of a parabola.

The Slowing of the Hands

After one-tenth of a second into the downswing, when the wrists are relaxed, the club is swinging out to hit the ball from the action of the centrifugal torque in the rotating system. From the curve for $\ddot{\alpha}$ we see that when this happens, the plot of the acceleration tends toward zero and then changes sign to become negative. A negative acceleration indicates that the arms and therefore the hands of the golfer are slowing down as the clubhead is approaching the ball. This effect, which has been observed from photographs, has been interpreted by some as indicating that "the golfer is quitting on the shot." But the slowing of the golfer's hands is a dynamic effect. In this part of the downswing the club has its greatest negative angular acceleration. The rotation of the club in the plane of the swing produces a force by the club on the golfer's hands that slows the hands even when the golfer is trying to move them with ever-increasing speed. This topic is discussed further in the Technical Appendix, Section 5.

The curve for $\dot{\alpha}$ shows a maximum positive angular velocity of the arms near the time when $\ddot{\alpha}$ is zero, while the curve for α shows an inflection at about this same time. The characteristics of these curves should be expected from the fact that $\dot{\alpha}$ is the time rate of change of α, and $\ddot{\alpha}$ is the time rate of change of $\dot{\alpha}$.

The discontinuity of the $\ddot{\alpha}$ curve is related to the abrupt change of the acceleration in the shift of the golfer during the downswing from a positive value through zero to a negative value. This abrupt change is a mathematical artifact of our model; in the real world these forces change gradually.

Wrist Torque During the Swing

Another quantity calculated by the computer is shown in Fig. 2.6. Here the torque, TH, which is defined in the Technical Appendix, Section 4, is the torque that the wrists of the golfer must overcome at the start of the downswing if the wrist-cock angle is to be maintained at a constant value. Thus the wrist must apply a torque TH to maintain a zero acceleration of the wrist-cock angle. In the first part of the swing a golfer would exert such a torque if he wished to maintain a constant wrist-cock angle temporarily. At the time when the curve of Fig. (2.6) goes through zero, he likely would exert no torque at all by his wrists on the club, allowing the torque TH alone to accelerate the uncocking process.

The stroboscopic photograph of the golf swing, which was analyzed to obtain the experimental data for the clubhead speed as it depends on time, could not provide data for the wrist-cock angle beta, the arm angle alpha, and the angular speed of the arms, as they depend on time.

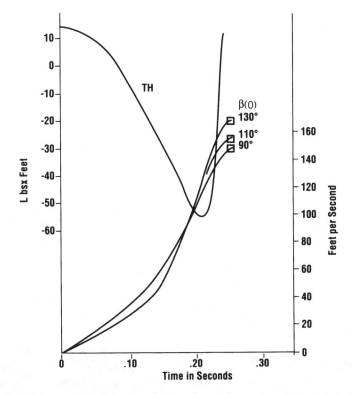

FIGURE 2.6. The curves show the clubhead speed as a function of time into the downswing for three different values of the initial wrist-cock angle. These curves show clearly that the clubhead speed at impact, dots enclosed in squares, increases as the initial wrist cock increases. This figure also shows the curve for the torque TH, which results from the inertial forces in the accelerating system acting on the club during the downswing. It begins with the positive value of about 15 lbs·ft and then falls to large negative values.

Since the experimental and calculated speeds of Tape A agree so closely, we have reason to expect that the calculated data for α, $\dot{\alpha}$, and $\ddot{\alpha}$ and β, $\dot{\beta}$, and $\ddot{\beta}$ would also agree closely with the same quantities if they could be obtained from the photograph. We may consider the swing characteristics shown in Figs. 2.4 and 2.5 to be similar to those of the actual swing of a golf club. Similarly, we have reason to expect that calculated data for torques within the system, such as TH, would also agree closely with the same quantities if they could be obtained somehow for the swing of the club that was photographed.

CHAPTER 3

A Second Look at the Golf Stroke

The Snapping of a Whip

The result of computer calculations of the swing may be met by considerable skepticism by many, and downright distrust by some, particularly the result that the torque on the club produced in the swing itself is very large and quite adequate to bring the clubhead to the ball at high speed without help from the golfer through the action of his wrists. The belief that wrist action by the golfer is a necessary part of the downswing is almost universal.

A friend and I were discussing this subject when he asked whether I had a club that he could swing to feel this large torque. We stepped out onto the patio with a driver. He took a few practice swings, and then told me he was going to swing the club keeping his wrists cocked throughout the swing. This he did. To his utter amazement the shaft broke in his hands near the lower part of the grip, and the clubhead and most of the shaft flew out over the garden. Later, I examined the broken shaft, but found no indication of any flaw in the metal. By not allowing his wrists to follow the club in its uncocking motion, the torque he developed was more than enough to break the shaft. I have not sacrificed another club to discover the size of this torque.

For the present, rather than look at further computer calculations, we shall take a second look at the golf stroke, using physical concepts in a qualitative fashion. What we find may be more easily accepted by golfers. We shall first compare the golf stroke to the snapping of a whip. The model appropriate to the study of a whip, such as a bullwhip, would be a large number of small flexibly connected rods rather than the two-rod model we have been using. The reason for looking at the bullwhip is that it is easier to see how physics is used to understand the whip action.

In pioneer days the bullwhip was a useful tool. Its main use was to make a loud noise, even though it could be used to produce damage. The noise is produced when the speed of the tip of the whip becomes greater than the speed of sound. The crack of a whip is thus similar in nature to the sonic boom produced by a supersonic plane.

Perhaps the reader is acquainted with this "crack-the-whip" action. As a child I played "crack-the-whip" at school recess, and my grandsons tell me that children still do it. In this activity, which is illustrated in a painting by Winslow Homer [5], children run together in a line, hand in hand, in the direction along the line. The leader stops suddenly, thus stopping the second child, and sequentially stopping the other children in a curving line. When this process comes to the end of the line, I know from my own experience that the last child is literally flying through the air. This action is similar to that of a bull-whip.

A bullwhip is made by braiding leather strips into a gradually tapering whip, weighted with lead shot at the thicker portion near the handle. Such a whip has a leather thong at its tip and may be as much as 12 ft or more in length. A high-quality bullwhip is surprisingly flexible.

In order to visualize what happens in the crack of a bullwhip, you should consider a very particular stroke of the whip. Start the stroke with the whip laid out on the ground behind you and extending back along a straight line with the tip farthest from you. You are then to swing your arm forward. This underhanded motion puts the whip in motion along a straight line, handle first, with the tip trailing along behind. At the end of the swing your hand stops and consequently the handle of the whip stops, but the rest of the whip continues to move in its original direction. Additional sections of the whip are sequentially stopped. The crack of the whip comes when all but the tip of the whip has been stopped.

How can physical principles help us to understand this action? At the start of the motion, as the hand moves the handle of the whip, the momentum of the whip increases. The hand exerts a force on the whip handle for a time, producing, according to Newton's second law, an increase of momentum. This force moving the whip handle a few feet also does work on the whip, giving it kinetic energy. When the hand stops, the whip exerts a force on the hand, and this force in turn decreases the momentum of the whip. Thus momentum is not conserved because a force acts. When the hand stops, the force of the whip on the hand does no work. There is a force, but the force is not accompanied by any displacement, since the hand remains at rest. If no work is done on the whip and no energy is lost in any other process, the kinetic energy of the whip remains constant. However, a small amount of energy is converted into heat energy.

During the stroke, successive parts of the whip are stopped, and the kinetic energy of these parts is fed into the successively smaller and smaller sections of the whip. The kinetic energy of a body depends on its mass and on the square of its speed. At the start of the stroke the total mass of the whip is moving with a moderate speed. Toward the end of the stroke, a much smaller mass must be moving at a much higher speed to have the same kinetic energy. For a tapered whip, even with some of the energy lost through internal friction, the speed of the very tip of the whip will reach a very high value. An experimental investigation [6] shows that the sonic boom interpretation of the crack of a bullwhip is correct; the fast-moving tip of the whip sends out a sound shock wave.

While we have not analyzed the action of a whip in complete detail, it is easy to see that the part of the whip that has been slowed and finally stopped has lost its kinetic energy, and the rest of the whip, the part still moving, has gained this energy. Similarly, in the golf swing, the arms, which have slowed in the latter part of the downswing, have lost kinetic energy while the club has gained kinetic energy. The dynamic process is the same in the two cases.

The energy put into the bullwhip in the form of work comes when the whip is pulled along its length by the action of the person swinging the whip. The pull is against the mass of the whip, giving this mass an acceleration. Is there a similar process in the swing of a golf club? How does the golfer put energy into the swing?

Large Muscles Supply Energy

When the large muscles of the body are suitably loaded, they are capable of producing mechanical power through a single contraction of about one-eight horsepower per pound [7]. I have made a rough estimate of the power of our professional golfer during the standard swing, and find that in part of the swing he delivers energy at about two horsepower. In the human body, for each muscle used to produce motion in one direction there is a muscle of about the same mass used to produce motion in the opposite direction. These considerations tell us that we should be looking for at least 32 lb of muscle to supply the power for the golf swing. If the muscles are not suitably loaded and hence cannot deliver one-eighth horsepower per pound, we should perhaps be looking for considerably more than 32 lb of muscle.

If you cannot visualize this much muscle, ask your butcher to put 32 lb of lean sirloin tip roast on the scales the next time you are buying meat. When you look for 32 lb of muscle on the average human body, you do not find it in the arms or even in the shoulders. You have to go to the legs, the thighs, and the back before you begin to approach this amount of muscle. Looking at the swing from the viewpoint of power shows in a convincing way that the energy of the swing must come in great measure from muscles other than those of the arms and shoulders.

Forces and Torques

We are thus faced with the problem of getting energy into the left arm of the (right-handed) golfer by having him use the muscles of his legs, thighs, and back. Not only must these large muscles supply energy, they must produce energy of rotation. I do not blame anyone whose first reaction is "It cannot be done. These large muscles are not even connected directly to the arms."

With the study of dynamics, a person develops some intuition concerning the type of motion that results when a particular body is acted upon by certain forces and torques. This intuition may be used in setting up mathematical statements involving Newton's laws of motion for exact descriptions of such motion. This is the process that was used in our analysis of the simple two-rod model of the downswing. Even without such an exact analysis, a person well acquainted with dynamics can foresee the general nature of the motion involved. The importance of the centrifugal torque in the motion of the club when it swings around to hit the ball can be realized without writing down any equations. The torque that is the reaction of the club on the hands, slowing them toward the bottom of the swing, is also apparent from the dynamics of the club motion. This intuitive approach may be used to understand further aspects of the golf stroke. The ideas presented here are of a tentative nature. Later, we shall bolster this approach by examining the differential equations and calculations relating to our two-rod model to see what they may tell us concerning the size of these torques and their timing.

Five Experiments

In the study of the downswing using the two-rod model, the torque acting on the golfer's arms is assumed to be given; its origin was not considered to be part of the problem. But the question of how the golfer's body is to produce the torques and forces involved in the downswing is central to the study of the golf stroke. To prepare ourselves to proceed with the analysis, we must look at the fundamentals of the motion of a rod when it is under the action of a force and a torque. The rod will represent the golfer's left arm, and for the present the golf club will be ignored.

The forces and torques acting during a vigorous golf stroke are so much larger than those produced by gravity that these latter forces and torques may be ignored in this consideration of the downswing. But when we take a rod in hand to study the motion produced when we apply various forces and torques to it, the effect of gravity interferes with the study. Perhaps our experiments should be performed by astronauts in an orbiting space capsule, where there is no gravitational acceleration relative to their surroundings. Here on the Earth's surface we have to cancel out the gravitational forces in some way. We can do this by laying a rod on a smooth surface of a level table. We wish to look at the effects of other forces on the rod. The frictional forces between the rod and the table cannot be eliminated, but happily, the frictional forces may be much smaller than the other forces involved. If the reader wishes to grasp the essentials of the motion of a rod, he should use a yardstick or similar object and watch its motion under the action of forces and torques when it is on a smooth table.

We shall examine five different types of motion of the yardstick and relate them to the motion of the golfer's left arm. Four of these motions are indicated in Fig. 3.1.

In our experiments the yardstick will represent the golfer's left arm. It will help to think of the golfer facing south as he takes his stance at the south edge of a table. We can then imagine that we tilt the golfer backward until the plane of his swing is level and then transfer the plane of the swing to the tabletop. An east-to-west line on the tabletop will represent the direction of the left arm when it is extended west horizontally in the backswing. A line on the table slanted to the north of the east-to-west line will represent the direction of the left arm when it is extended above the horizontal in the backswing. A north-to-south line on the table will represent the direction of the left arm when it is extended toward the ball at the bottom of the downswing. The motion of the left arm in the downswing, as seen from above the table, is represented by a counterclockwise rotation.

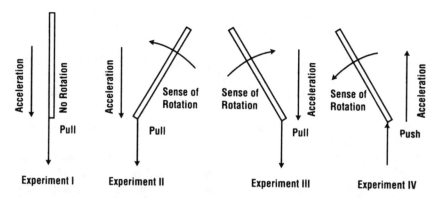

FIGURE 3.1. Diagrams that may be used to help understand how the shift of a golfer is able to produce rotational acceleration of the golfer's arms as explained in the text.

The First Experiment

Place the yardstick on a smooth table along an east-to-west direction and pull on it directly to the east by means of a short string attached to its east end. Observe that the yardstick accelerates in the direction of the force on it but does not show any rotational motion. This result tells us that if the golfer pulls horizontally on his left arm by his left shoulder while the arm is in a horizontal position, the arm will not be set into rotational motion by the pull, although it will have a linear acceleration to the left. For the left arm in this position, any rotational motion produced must come either from a torque on the arm by the shoulder muscles or by gravity. For this discussion we shall call this linear acceleration positive.

The Second Experiment

Place the yardstick on the table again but with the yardstick slanted to the north of the east-to-west direction. Pull the string to the east and observe that the yardstick accelerates toward the east and starts to rotate at the same time. It has an angular acceleration. The rotation of the yardstick brings it toward the east-to-west direction. This result tells us that when a golfer has his left arm above the horizontal at the top of the backswing and pulls horizontally on the arm with his left shoulder, the arm will have a linear acceleration to the left and an angular acceleration in the sense to start the arm moving in the down-swing. We shall call this a positive angular acceleration for this discussion. This positive angular acceleration is not produced by the shoulder muscles or by gravity; it is produced solely by the horizontal pull by the left shoulder. You should do this experiment with various angles of slant to the north of the east-to-west direction so that you may observe that the angular acceleration increases with the angle of slant.

The Third Experiment

Place the yardstick on the table but now with the yardstick slanted to the south of the east-to-west direction. Pull on the string toward the east and observe that the yardstick accelerates toward the east and starts to rotate at the same time very much as in the previous experiment, but now the yardstick starts to rotate in the opposite sense; it has a negative angular acceleration. This result tells us that if the golfer pulls horizontally on his left arm by his left shoulder while the arm is extended below the horizontal, the angular acceleration will be in the sense to lift the left arm if it is originally at rest and to slow up any angular motion the arm may have if it is moving in the downswing. The arm will have a negative angular acceleration and a positive linear acceleration.

The Fourth Experiment

Place the yardstick just as in the previous experiment, but now the pull on the string is to be replaced by a push toward the west at the east end of the yardstick. The yardstick will now have a linear acceleration in the opposite direction from that in the previous experiment. The angular acceleration will be in a sense to increase the slant of the yardstick to the south away from the east-to-west direction. The angular acceleration will thus be positive, as it was in the second experiment. This result tells us that if the golfer no longer pulls horizontally on his left arm by his left shoulder when the arm is extended below the horizontal but rather slows up the motion of the shoulder, giving the arms a negative linear acceleration toward the west, the arm will have a positive angular acceleration in the sense to speed up any rotational motion the arm may have in the downswing.

The Fifth Experiment

Place the yardstick on the table in an east-to-west position and grasp the east end between the fingers and give the yardstick a counterclockwise torque as seen from above the table about a vertical axis at your fingers. Observe that the yardstick will move with a positive angular acceleration about the chosen axis. This result tells us that such a torque on the golfer's left arm by his shoulder and back muscles will give the arm a positive angular acceleration to speed up its angular motion in the downswing.

In the discussion of these five experiments we have omitted any reference to the fact that the golfer's arm is connected to the golf club. When the golfer pulls on his left arm with his left shoulder, he really pulls against the mass of the left arm and the golf club. In our discussion we shall call the left arm and the golf club the "system." The center of mass of this system is always above that of the left arm alone; the line on the table representing the golfer's left arm should be replaced in our thinking by a line to the center of mass of the system. The general conclusions of the previous discussion remain valid with this change.

Left Shoulder Pulls on the Left Arm

With these experiments as background, we are now ready to make suggestions as to how a golfer should organize the downswing of his golf stroke so that at no time during the stroke will he be doing anything to hinder the generation of the needed power. We have seen that the horizontal pull by the left shoulder on the left arm will produce a positive angular acceleration to help with the downswing only when the system extends above the horizontal (experiment II, Fig. 3.1). The large muscles should do their work in the early part of the downswing, before the system has reached the horizontal. If the left shoulder continues to pull on the system after it has passed the horizontal, the negative angular acceleration produced actually slows the downswing (experiment III). These observations indicate quite clearly that the downswing should not be started by the sole action of the shoulder and back muscles in producing a positive torque on the left arm with the action of the large muscles coming later, but rather that the large muscles of the arms, shoulders, and back should act together.

Organization of the Swing

Our five experiments lead us to the view that the downswing should be organized with a major torque (TS) produced by the shoulder and back muscles of the golfer and with torques produced by the shift of the golfer toward the target using the large muscles of his hips and legs. These latter torques come about first through the pull of the left shoulder on the near end of the system (arm and club) and later, when the center of mass of the system is a little below the horizontal, through the push of the left shoulder on the near end of the system. Both of these actions, the pull and push, produce torques that help in the downswing. These actions should be such that the pull and push are completed by the time the ball is hit. By that time the left shoulder should have essentially finished its lateral motion.

The starting and stopping motion of the left shoulder must take place over a very limited distance. From photographs of competent golfers I have estimated this distance to be usually about 15 inches. This distance is along a somewhat curved path of the left shoulder. Some golfers may extend this distance to as much as 20 inches.

Let us look at what the calculations of the standard swing can tell us about how much of the clubhead speed at impact results from the acceleration of the system produced by the pull and push on the left shoulder. The results of these calculations, while given in exact numbers, should be interpreted only in a qualitative sense. A major reason to be skeptical of a quantitative interpretation stems in part from the difficulty of determining the mechanical properties of the golfer's arms. The increment to the clubhead speed at impact resulting from the action of the left shoulder is almost directly proportional to the first moment of the golfer's arms, which can be determined only by a rough estimate.

A good match between the simple model and the data from the stroboscopic photograph could be obtained with the following maneuver. First, a pull on the left arm produces an acceleration of 47.4 ft per second per second (ft/s^2) for the first 0.16 s of the downswing, while the golfer's arm is above the horizontal, and then a push produces a negative acceleration of 50.6 ft/s^2 beginning 0.17 s into the downswing. Remember the acceleration of gravity is 32.2 ft/s^2 each second. The shift velocity when the ball is hit was 3.3 ft/s, and the shift when the ball was hit was about 14 in. The maximum shift velocity during the downswing was 7.6 ft/s. This shift in the standard swing gives a clubhead speed at impact of 163.4 ft/s. A calculation of the standard swing with no shift whatever gives a clubhead speed at impact of 149.0 ft/s. We may conclude that the use of this shift in the swing gives an increase in clubhead speed at impact of 9.7%.

The golfer has great latitude in choosing how he is to apply the force on the upper end of his left arm. It is of interest to use the standard swing with various positive and negative accelerations to see whether a larger effect on the clubhead speed than that shown in the previous paragraph can be obtained. The shift previously described is obviously faulty because the negative acceleration is not large enough to stop the shift by the time the ball is hit. A shift calculated with this fault corrected gives a clubhead speed at impact of 170.5 ft/s. This shift had accelerations of 67.1 and −92.6 ft/s^2 acting for 0.165 and 0.085 s, respectively. The total shift was 15 inches in this swing. Correcting this fault increases the clubhead speed at impact to 14.4% over the speed for the swing with no shift.

The problem of what shift action the golfer should use in the standard swing shall be left unsolved. Many different shifts were calculated, but none gave a clubhead speed greater than 174.5 ft per second. This shift, possibly the optimum shift, gives a clubhead speed 17.1% greater than the swing with no shift. Swings in which the shifts came earlier and later in the swing, with coasting times with no acceleration in the middle of the swing when the golfer's arms were nearly horizontal, gave clubhead speeds less than that of the possible optimum shift.

The increase in clubhead speed when the golfer uses a shift compared with the clubhead speed with no shift is not great. However, playing golf without using a well-constructed shift would put a golfer at a troublesome disadvantage.

There seems to be no discussion in the golf literature that presents an analysis similar to our second look at the golf stroke. This analysis, which is not complete, is based on fundamental physical principles and follows my own intuitive approach in their application. The skeptical reader may be reluctant to consider the organization of the downswing suggested here as having any relation to what the competent golfer actually does when he swings a club. There seem to be no experimental studies that tell us exactly how the force between the golfer's left shoulder and his left arm varies during the downswing. However, we may find it of interest to look at remarks made by expert golfers in their subjective descriptions of their downswings to see whether we find any contradiction to the organization of the stroke suggested here.

The Word of Experts

On returning to the various books by the experts, I was surprised to find how little has been written on the details of the downswing. Bobby Jones [8] tells us that a golfer necessarily plays by feel, and feel is almost impossible to describe. He does tell us that when he was playing well he had a feeling of pulling against something. He does not elaborate on this point, but it is reasonable to identify this pull with the action of the large muscles of the body, as suggested earlier. He also mentions a definite quick shift to the left at the beginning of the downswing. Jack Nicklaus [9] also refers to this quick shift to the left and tells us that the quicker this shift, the more powerful the stroke. We can look upon the shift as the result of the early action by the large muscles of the body.

The suggested delay in the action of the shoulder muscles can be interpreted subjectively as a slow start on the downswing. We find many experts who insist that a slow start on the downswing is necessary in a good golf stroke. Sam Snead [10] advises the golfer to accelerate the downswing gradually. Tommy Armour [11] tells us that we cannot get power into a stroke if we are hasty starting down. Cary Middlecoff [12] remarks that a fast start on the downswing is the great swing-wrecker of golf. Frank Beard [13] wants to delay any hand action until the stroke is well started. All of these remarks, although they do not refer directly to a delay in the use of the shoulder muscles, are general enough so that they may be interpreted as supporting our suggested organization of the downswing.

The torque applied by the shoulder muscles is not discussed as such in golf literature, but a number of writers refer, as Frank Beard does, to hand action. As I understand it, this hand action is the bringing of the hands along the arc they must follow in the downswing. Their motion results from the action of the shoulder muscles and not from the use of the forearm muscles. Several of the experts recognize that this bringing of the hands into the stroke comes only after the downswing is well started.

The action of the large muscles of the body in decelerating the left shoulder is also not discussed as such in golf literature, but the subjective side of this process is described by the well-known phrase "hitting into a firm left side." Al Geiberger [14] writes on the importance of a firm left side. Jack Nicklaus [9] gives an excellent description of this process when he tells of the stopping of the swift lateral action of the body to release the energy of the downswing. Julius Boros [15] wants his body to be almost motionless at impact.

Timing

Many experts have expressed subjective descriptions of their downswings that point to actions in agreement with the organization of the downswing suggested. The application of physical principles explains why the action of the large muscles of the body must start the downswing, why the maximum action of the shoulder muscles may be delayed somewhat until the action of the large muscles is well underway, why the shoulder muscles must then act continuously until the ball is on its way, and why the large muscles of the body act again toward the end of the downswing to slow up the body motion to give the system (arm and club) additional positive angular acceleration after it has passed the horizontal in the downswing. We find that these various actions must be performed in a definite sequence and at appropriate times. Perhaps the sequence of actions at appropriate times is what is implied by the nebulous descriptive term "timing" when applied to the golf stroke. Bobby Jones wrote that we talk about good timing and faulty timing, and the importance of timing, and yet no one has been quite able to say just what timing is. By the application of physical principles we have come to understand, at least in some measure, the timing of the various actions in the downswing. I suggest that the sequence of events previously described constitutes good timing, while a stroke with the hips moving simply to "get the hips out of the way of the hands," a sequence recommended by some writers, would be one with seriously faulty timing.

A perusal of the detailed study of the five torques acting on the arms and the five torques acting on the club during the downswing, presented in the Technical Appendix, Section 5, may give the interested reader an appreciation of the complexity of what happens in the downswing of the golf stroke.

A golfer who is not convinced of the efficacy of the shift of the center of rotation during the downswing should not consider himself alone in this thinking. Many sources advocate swinging the club about a quiet center with little head motion and even suggest that someone should hold a hand on a golfer's head during a downswing in the study of head motion. While a quiet head does not necessarily mean that there is no motion of the center of rotation during the downswing, the motion of the head and the center of rotation are not usually clearly distinguished.

To emphasize the thinking concerning a quiet center of rotation during the downswing and to leave no doubt that this point of view indeed appears in the golf literature, I shall quote a passage from the *Search for the Perfect Swing* [7].

In discussing the downswing the authors write,

> The other main point to stress is that he (the golfer) must keep the center of the whole action firmly fixed, as the central point is fixed in the model, if the swing is to work in the simplest and most powerful form, which alone can make it a consistent and repeating one. He must do so from at least the beginning of the forward swing up to the moment of striking the ball.

> This is a fact that can be derived from the mechanical model. A golf swing can work effectively, after a fashion, even if the central pivot is moved forward toward the hole during the downswing—so long as the whole action of the two-lever system is timed to flow with that movement of the pivot. But moving the pivot (which, in effect, means the whole body) during the forward swing inevitably reduces the player's ability to generate the greatest possible clubhead speed into impact, for the very simple reason that it uses up, wastefully, energy which might have gone into the clubhead. It also makes the swing's working more complicated in a way which most men and women could very well do without.

The point of view expressed by these authors seems to indicate that they have failed to take into account the physical reason for the use of the shift during the downswing of a golf club. The properly timed shift produces torques by the large muscles of the body that add to the energy of the clubhead at impact with the ball and does not reduce the energy produced by the golfer in the downswing.

Shape of Path of Clubhead Confirms Shift

It is of interest to consider the shape of the path of the clubhead during the downswing as supporting the presence of the shift during the swing of the professional golfer who swung the club for the stroboscopic photograph used in this study.

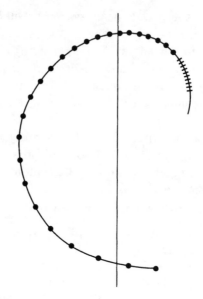

FIGURE 3.2. Trace of the clubhead positions from an enlargement of the photograph shown in Fig. 2.1.

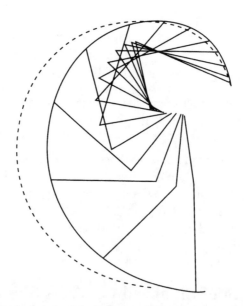

FIGURE 3.3. The calculated stick diagram of the positions of rod A (arms) and rod C (club) for equal time intervals during the standard swing. The total shift of the pivot assumed in this case was about 13 in. The continuous curve shows the positions of the clubhead during the downswing. The dashed curve shows the calculated positions of the clubhead during the same downswing when the shift of the pivot was subtracted from the clubhead positions with the shift.

The criterion used in obtaining the calculation of the standard swing is that the velocity of the clubhead in the calculation should match the velocity of the clubhead obtained from the stroboscopic photograph. The shape of the clubhead path was not used in the determination of the standard swing.

The photograph in Fig. 2.1 was enlarged to a convenient size. A trace of the clubhead positions during the downswing was obtained from this enlargement and is shown in Fig. 3.2.

The calculation of the standard swing was used to produce a stick drawing of the positions of rod A and rod C for equal time intervals during the downswing (Fig. 3.3). The size of this drawing was chosen to match that of the enlarged photograph. A continuous curve was drawn showing the positions of the clubhead during the downswing. This curve is traced and presented in Fig. 3.4 by means of crosses. A trace from Fig. 3.2 is presented in Fig. 3.4 by means of dots. The agreement of the curves shown by the crosses and the dots is striking. The difference between these two traces toward the end of the downswing comes about because the shift assumed in the calculation is along a

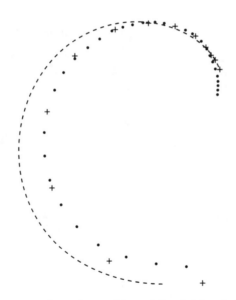

FIGURE 3.4. The crosses show the position of the clubhead at equal time intervals obtained from the calculation of the standard swing; the dots show the positions of the clubhead during the downswing at equal time intervals from the stroboscopic photograph of the swing of the golfer; and the dashed curve shows the position the clubhead would have during the downswing from the calculation of the standard swing with the horizontal displacement of the pivot subtracted out. The first two curves agree throughout most of the swing. The dashed curve does not fit the data from the standard swing or the data from the stroboscopic photograph.

horizontal straight line, while the shift of the golfer is probably along a line that begins to curve upward just before the ball is hit.

The stick drawing in Fig. 3.3 may be modified by removing the shift for each position of rod A and rod C. The dashed curve in Fig. 3.3 shows such a clubhead path for the standard swing with the shift removed. A trace of this curve is also shown in Fig. 3.4. The difference between the clubhead paths of the stick drawings of the standard swing and of the photograph, and the clubhead path of the standard swing without the shift, is very conspicuous.

This study of the path of the clubhead during the downswing shows beyond any doubt that this professional golfer, known for his distance off the tee, uses the shift to produce extra torque that yields high clubhead speed during the downswing. I doubt that he is the only golfer who uses the shift to achieve extra clubhead speed.

Chapter 4

Variation of Parameters Brings New Understanding of the Golf Swing

The Variation of Parameters

We shall now look at what happens in calculated swings that are somewhat different from the standard swing. As an example we shall study swings that result when the golfer uses shorter or longer backswings.

I once watched a university student hitting balls with his driver in the university practice cage. He had the largest backswing I had ever seen. When I asked him whether he drove the ball a great distance, he proudly replied that indeed he did. I hazarded another question: "Yes, but you do not know where the ball is going to go, do you?" He replied: "That is what the coach keeps telling me." One professional golfer told me that in his coaching he finds it most difficult to persuade a golfer to shorten his backswing. It is generally known that a large backswing may be used to produce long drives, but it is also generally known that a large backswing produces an adverse effect on the precision of the shot. Presently we shall show that a backswing may be reduced a considerable amount without more than a slight decrease in clubhead speed at impact with the ball. Knowledgeable golfers know this to be true. Understanding this effect should relieve the golfer of any anxiety in considering a shortened backswing.

We shall attempt to extend our understanding of the actual swing of a club by examining what happens to the clubhead speed in calculations, according to our model, when parameters are varied, one or two at a time, from the values used in the calculations for the standard swing. Unless otherwise indicated, we shall maintain all other parameters of the swing unchanged. The results of these calculations will be presented as actual numerical values, but they should not be considered as applying quantitatively to the actual swing of a club. Remember, we are using a simplified model of the golf swing.

33

Torque by Arms

An obvious way for an aspiring golfer to increase his clubhead speed is somehow to increase the torque TS he is able to apply to the rotating system during the downswing. Some have advocated that the golfer enter into a physical education program to increase his muscular strength. If a golfer were to increase this torque by some factor, would the clubhead speed increase by the same factor?

One way to find the answer to this question is to increase the parameter TS in the calculation of the standard swing by 5% and find the effect on the clubhead speed at impact with the ball. The result of this calculation is somewhat disappointing. The clubhead speed at impact increases by only 1.7%. Thus an increase in the torque TS does not produce a proportional increase in clubhead speed. The value of $\beta(i)$ decreases by 0.45 deg.

A 5% decrease in the torque TS produces only a 1.8% decrease in clubhead speed at impact with the ball. The value of $\beta(i)$ increases by 0.53 deg. This is what is behind the advice given by some accomplished golfers who advocate swinging the club at somewhat less than full power. Sam Snead [10] says that you must learn to swing within yourself, and he estimates that 85% of full power is enough.

The Effect of Gravity and Shift

Another calculation was made to determine the effect of the force of gravity on clubhead speed in the standard swing. In this calculation the value of the parameter g, the acceleration of gravity, was set equal to zero. The effect of this change was to reduce the clubhead speed at impact with the ball by 8.5% and to decrease $\beta(i)$ by 2.9 deg. A similar calculation using the standard swing with no shift of the golfer toward the target showed an 8.8% decrease in the clubhead speed at impact with the ball and an increase of $\beta(i)$ of 4.2 deg.

Wrist Cock at the Top of the Backswing

What effect does the wrist-cock angle at the beginning of the downswing have on the clubhead speed at impact with the ball? We shall seek the answer to this important question by making the calculation for the standard swing for various different values of the parameter $\beta(0)$, the wrist-cock angle at the beginning of the downswing. Preliminary calculations showed that in order to hit the ball with the standard $\beta(i)$ of 15.3 deg, the backswing angle had to be adjusted for each value of the wrist-cock angle $\beta(0)$; larger values of $\beta(0)$ necessitated larger values of γ. Calculations were made for wrist-cock angles at the start of the downswing of 90, 110, and 130 deg. The value of γ for each swing was adjusted to ensure the same wrist-cock angle $\beta(i)$ at impact. The complete curves for the clubhead speeds for $\beta(0)$ of 90 and 130 deg are shown in Fig. 2.6.

The different initial slopes of these curves are related to the different distances of the clubheads from the axis of rotation at the start of the downswing. The 90 deg value of $\beta(0)$ puts the clubhead at a greater radius at the start of the downswing than for the larger values of $\beta(0)$, and for a similar angular acceleration the clubhead moves with a greater linear acceleration. The upper segment of the clubhead speed curve for $\beta(0)$ of 110 deg is also shown in Fig. 2.6. The main point of interest here is that an increase in the wrist-cock angle at the beginning of the downswing produces an increase in the clubhead speed at impact with the ball. The backswing angle must be successively increased as $\beta(0)$ is increased to maintain an unchanged wrist-cock angle at impact. The increase in the clubhead speed at impact may be attributed to the fact that the torque TS by the golfer's arms acts through a larger angle before the ball is hit.

Reduced Backswing

Let us look again at the clubhead speed curve in Fig. 2.2. In this curve the wrist-cock angles throughout the swing are indicated in multiples of 10 deg by small circles. Since the clubhead speed curve shows a flattening off toward the end of the swing, the clubhead is seen to approach a nearly constant speed about the time the ball would be hit with a zero wrist-cock angle. If we look at the numbers along the curve indicating the angle of the wrist cock at various clubhead speeds, we find that the clubhead speeds vary relatively little while the wrists are uncocking through their last 30 deg or so. From the curve in Fig. 2.2 for the standard swing we find that the clubhead speed for a β of 30 deg is about 3.9% less than for a clubhead for β equal to zero. The backswing angle for the standard swing is 165.8 deg, and for a swing that would allow the golfer to hit the ball with a zero wrist-cock angle, the backswing angle would have to be increased to 185.8 deg. The respective clubhead speeds on impact for these two swings are 163.4 and 167.6 ft per second. The golfer may therefore reduce his backswing angle 30 deg and hit the ball with a wrist cock angle of about 15.3 deg and lose about 2.5% of possible clubhead speed. We may, according to our calculations, make the general statement that a golfer may reduce his backswing angle substantially and, with the same torque, achieve almost the same clubhead speed at the impact with the ball as with a much larger backswing. If he hits the ball this way, he hits with his hands well ahead of the ball at impact, or with what is known as a late hit.

The calculations show this to be true, but does any golfer in his right mind hit the ball with his hands ahead of the ball to any extent? Our professional golfer does. There are a few high-speed photographs in the golf literature showing expert golfers caught with the clubhead in contact with the ball, and in all of them the wrist-cock angle is substantially larger than zero. These golfers certainly take advantage of the physical nature of the golf swing to achieve the greatest clubhead speed without the loss of precision likely with an exaggerated backswing.

The Effect of Wrist Action

The wrist action of a golfer is of greatest importance in all his swings. We have seen the effect of the size of the wrist-cock angle at the beginning of the downswing on the clubhead speed at impact with the ball for swings that are otherwise the same except for the necessary adjustments in the backswing angle. We may continue our study by looking at the effect on the swing that results from other action the golfer may take with his wrists during the downswing.

At the start of the downswing the golfer has cocked his wrists at some angle $\beta(0)$. If the golfer were to begin the swing with flexible wrists and keep them flexible, much to his surprise, the golf club would swing in and probably hit him in the back of his neck.

This peculiar action can be demonstrated very easily. The reader may balance a club between the thumb and first finger of his right hand holding the club and his forearm in a horizontal position with his elbow against his side. He is then to move his hand in a horizontal circle with his elbow acting as the center of the circle. If the club is held loosely, it will keep its direction in space, and the angle between the forearm and the club will change in such a manner to account for the motion previously mentioned. If the club is held tightly between the thumb and first finger, a torque will be felt when the rotary motion is repeated with the angle between the club and the forearm unchanged. The action of the club in maintaining its direction in space unless a torque acts on it is an example of the inertia effect described in Newton's first law of motion. This simple experiment is of value in giving the golfer a feeling for what happens at the beginning of the downswing.

This demonstration shows that if we are to maintain the initial wrist-cock angle $\beta(0)$ during the start of the downswing, the wrists must exert a torque on the shaft of the club in the sense that would decrease the wrist-cock angle. This torque by the wrists must be equal and opposite to the inertial torque *TH* that we have plotted in Fig. 2.6. This inertial torque, first positive, gradually decreases to zero and then becomes negative. In the standard swing the torque by the wrists must therefore counter the inertial torque and begin with a negative value of about 15 lbs-ft and gradually decrease to zero at about one-tenth of a second into the downswing.

The Uncocking of the Wrists

After one-tenth of a second, the inertial torque *TH* continues to act on the club, and the clubhead moves out to hit the ball, helped or hindered by any further wrist action by the golfer. In the calculation of the standard swing, it was assumed that the golfer maintains the original wrist-cock angle $\beta(0)$ until the torque by the wrists decreases to zero, after which the wrists of the golfer offer no help or hindrance until the clubhead hits the ball. Thus the action of the golfer's wrists, after the torque *TH* decreases through zero, is entirely passive. Bobby Jones has written that it feels like the club is "freewheeling" through the ball.

The golfer may modify the wrist action described in the previous paragraph by applying an additional torque, negative when it helps in the uncocking process and positive when it hinders it. We may study the effects of this additional torque by putting a torque *TE* (*E* for extra) into our calculation of a thus modified standard swing.

If a golfer takes a driver and holds it at rest horizontally by the grip, the torque he feels is almost 2 lbs-ft. Let us look at two calculations with the standard swing modified by an additional uncocking torque of 2 lbs-ft lasting for 0.10 s from the start of the downswing, with the wrists entirely passive for the rest of the swing. One calculation will have the backswing angle γ decreased somewhat, and the other will have the same angle γ as for the standard swing.

In the first of these calculations the backswing angle is reduced from 165.8 deg to 154.6 deg so that the wrist-cock angle when the ball is hit is the same 15.3 deg as in the standard swing. In the second calculation the backswing angle is kept at 165.8 deg, and the wrist cock angle at impact is found to be 6.9 deg. For the first calculation the clubhead speed on impact with the ball is 5.0% less than for the standard swing, and for the second it is 2.9% less than for the standard swing. The calculations show that the clubhead speeds for the two modified swings are greater than the clubhead speeds of the standard swing for most of the swing, but toward the end of the swings the clubhead speeds become less than the speeds of the standard swing.

The Paradox

If a golfer were to use wrist action in the way we have described to bring the club around to hit the ball, he would undoubtedly expect increased clubhead speed at impact with the ball. But the calculations show that this does not happen. The results of these calculations are truly surprising. We are tempted to conclude, from what we have seen so far, that anything the golfer does with his wrists during the swing to decrease the wrist-cock angle will result in less clubhead speed when the clubhead meets the ball. Could this be a general guiding principle for the swing of a golf club?

This whole idea appears so paradoxical that many golfers would reject it out of hand. Perhaps the decrease in clubhead speed comes about because the helping torque is applied for such a short time at the beginning of the downswing. Let us consider a calculation having a helping torque by the wrists of −4 lbs-ft, twice the former value, and have the torque act continuously throughout the downswing. Certainly, when the golfer puts such effort into the swing throughout the whole downswing, the clubhead can be made to move faster when it hits the ball.

Such a calculation was made with all the parameters of the standard swing unchanged and with TE of −4 lbs-ft used throughout the swing. The backswing angle was unchanged, and for that reason the wrist-cock angle $\beta(i)$ on impact came out to be −15.2 deg instead of the +15.3 deg of the standard swing. The negative value of $\beta(i)$ means that on impact with the ball, the clubhead is ahead of the hands. The helping torque of −4 lbs-ft by the wrists brings the clubhead around much too quickly.

Again the calculations show that for most of the swing the clubhead speed is greater than the speed of the standard swing. However, the ball is hit with a clubhead speed 5.7% less than the speed of the standard swing. We find indeed that through this wrist action the golfer does not produce the expected clubhead speed on impact, and this is true no matter how hard he tries. A slightly different example will give an additional reason to believe that anything the golfer does to reduce the wrist-cock angle during the downswing will give a smaller clubhead speed at impact.

Loose Grip and Bent Elbow

Many golfers bend the left arm or loosen the grip at the top of the backswing; some do both. At the beginning of the downswing these golfers will straighten the arm, tighten the grip, or both.

These acts will give the club a small angular velocity, reducing the wrist-cock angle. Let us look at a calculation for the standard swing with all the parameters unchanged but with a small angular speed of wrist uncocking at the start of the downswing. We shall arbitrarily choose $\dot{\beta}(0)$ to be -2 rad per second. A body with this angular speed will be turning around once in about 3 seconds, and the greatest angular speed of the club in the standard swing is about -30 rad per second. This angular speed gives the clubhead a running start in the downswing of about 6 ft per second. Except for this initial speed, the curve of clubhead speed is practically the same as that of the previous calculation. The ball is hit with a clubhead speed 5.3% less than that for the standard swing and now with a wrist-cock angle at impact of only -0.5 deg.

In the three calculations just considered, the golfer acts early in the downswing to help in the uncocking of the wrists. In each of them this action results in a decrease in the clubhead speed at impact compared with that of the standard swing. Let us look at the effect of such a helping action of the wrists late in the downswing. Extensive calculations with the standard swing modified by a constant helping torque of 2 lbs-ft by the wrists late in the downswing show that indeed such a helping torque does produce an increase in clubhead speed at impact. The maximum increase in clubhead speed at impact comes with the torque starting about seven-hundredths of a second before the ball is hit. The resulting clubhead speed was found to be only about 0.7% greater than that of the standard swing. This constant torque acting for a longer time or for a shorter time at the end of the downswing produces clubhead speeds less than the maximum.

From these calculations it appears that any helping torque early in the downswing will have an adverse effect on the clubhead speed at impact, while such a torque late in the swing, with proper timing, may be desirable if the golfer is able to apply it without an adverse effect on the precision of the swing. With Bobby Jones's freewheeling through the ball, there is no such wrist action that might alter the path of the clubhead as it approaches the ball. For those who wish to try to build this late wrist action into their swings, it may be of interest to know that for the standard swing this wrist action must be delayed until the arms are about 60 deg back from the vertical.

Is Hindering Wrist Action Practical?

The adverse effect of any wrist action early in the swing to help the uncocking raises the question of whether a hindering torque early in the swing will produce the opposite effect. A calculation was made with the standard swing modified by having the wrists hinder the uncocking process by a torque of 2 lbs-ft acting for 0.10 s. The backswing angle was increased from 165.8 deg to 173.6 deg so that the wrist-cock angle at impact is the same 15.3 deg as for the standard swing. The clubhead speed on impact was found to be 5.4% greater than that for the standard swing. If the same wrist action is applied without the increase in the backswing angle, the clubhead speed at impact was found to be only 2.9% greater than that for the standard swing, and the angle $\beta(i)$ at impact was found to be 26.3 deg.

Bobby Jones writes that at the beginning of the downswing one should have the feeling of leaving the clubhead at the top. We might interpret this statement to mean that the positive angular speed of cocking of the wrists is desirable at the beginning of the downswing. This action of the wrists would be the opposite of the adverse effect in the swing of a golfer with a loose grip or a bent left arm. Numerous calculations to explore this area show that with an initial continuing cocking motion of the wrists at the beginning of the downswing, a $\dot{\beta}(0)$ of plus 2 rad per second as an example, an otherwise standard swing would indeed give a slight increase in clubhead speed at impact, but with a wrist-cock angle at impact that is probably too large to be practical. Increasing the backswing angle and decreasing the initial wrist-cock angle so that the wrist-cock angle at impact $\beta(i)$ has a reasonable value for such a swing does not give a clubhead speed at impact sufficiently greater than that of the standard swing to suggest that such wrist action is of any value. The feeling of "leaving the club at the top" at the start of the downswing should probably be interpreted as consciousness of not doing anything to produce an early uncocking of the wrists.

Substantial Wrist Cock at Impact

One characteristic of all swings of competent golfers, for clubs from drivers to wedges, is that the clubhead meets the ball when there is a substantial wrist-cock angle. While there is no uncertainty in the wrist-cock angle at impact in our model, it is difficult to specify unique wrist-cock angles for golfers from most of the photographs one finds in the golf literature. We may draw a line from the clubhead to the golfer's wrists with little error, but a line from the point about which the wrists cock to the center of rotation of the swing is another matter. We may arbitrarily choose a point midway between the golfer's shoulders as the center of rotation and draw a line from that point to the point 5 in from the grip end of the club. The angle between these lines allows at least rough comparisons between wrist-cock angles near impact for photographs of swings showing the clubhead close to the ball. Wrist-cock angles at impact so determined are generally in the range of 20 to 45 deg.

CHAPTER 5

The Energy of the Swing

Sources of Energy

In the early days of golf, the golf swing could not have possibly been considered in terms of the concepts used in this chapter. The concepts of work and energy, as used in mechanics, did not exist in the 15th century when golf was being played in Scotland. The words "energy" and "work" first appeared in the present context in 1807 and 1826, but the ideas behind these words developed earlier. It took about 100 years from the first glimmer of these ideas before we find the statement of the principle of conservation of energy in 1847. This principle as applied to all forms of energy is one of the fundamental ideas of physics. The ancients knew that hands rubbed together became warm, but this effect was one of the mysteries.

In the development of our understanding of the dynamics of the swing of a golf club we have used the concepts of force, torque, and linear and angular motion. We may further this understanding if we look at the swing of a club using the concepts of work and energy.

In the golf stroke some of the energy supplied by the muscles of the golfer and the energy coming from the fall of the arms and club in the gravitational field is transferred to the golf ball at impact. Since the transfer of energy from the clubhead to the ball during the collision between them is determined by the physical properties of the clubhead and the ball and by the laws of physics, there is nothing the golfer can do during the collision to influence this transfer in any way. We may therefore focus our attention on how the club acquires its energy during the downswing.

In our two-rod model of the downswing, the energy put into the moving system comes from three sources. The major source is the work done by the golfer when the torque TS acting on the arms moves them through the angle of the downswing. A second source of energy is the potential energy of the arms and club at the top of the backswing; this potential energy changes into kinetic energy as the arms and club fall during the swing. The third source of energy is the work done by the golfer during the shift of the axis of the swing.

The Calculation of Energies

A computer was programmed to give, for the standard swing, the total kinetic energy of the arms and club and that of each of them separately. These energies are plotted in Fig. 5.1 in Curves A, B, and C as functions of the downswing angle.

Let us first consider Curve D in Fig. 5.1. This curve shows the work done by the golfer as a function of the downswing angle as he applies the torque TS on his arms. Since the work he does is the product of the constant torque TS and the downswing angle α, it is then directly proportional to the downswing angle measured from the beginning of the downswing. This curve is therefore a straight line. This curve also represents the contribution to the kinetic energy of the arms and club from this work done by the golfer. If we add to this curve the work done by the golfer in his shift toward the target and the kinetic energy coming from the decrease in potential energy previously mentioned, we arrive at the total kinetic energy of the arms and club shown in Curve A. This curve is not a straight line, since the contribution to the total kinetic energy of the system by the decrease in the potential

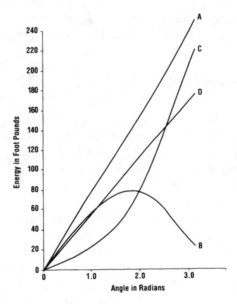

FIGURE 5.1. How energies of the system vary with the angle α into the downswing. Curve D, a straight line, shows the work done by the golfer as he applies the torque by his arms to the system. Curve B shows how the kinetic energy of the arms varies throughout the downswing. Curve C shows how the kinetic energy of the club alone varies into the downswing. Curve A shows the total kinetic energy of the system as it develops throughout the downswing.

energy of the arms and the club and the work done as the golfer shifts toward the target is not a linear function of the downswing angle. The total kinetic energy of the system when the clubhead makes contact with the ball comes 71% from the work TS × $\alpha(i)$, 13% from the decrease in the potential energy of the system, and 16% from the work done on the system in the shift of the golfer toward the target.

The Kinetic Energies

When we calculate the kinetic energy of the arms alone, we obtain the results shown in Curve B in Fig. 5.1. This curve shows that the kinetic energy of the arms reaches a maximum and then decreases. This decrease is related to the slowing of the hands late in the downswing. When we calculate the kinetic energy of the club alone we obtain the results shown in Curve C. The sum of these two kinetic energies is the total kinetic energy of the system and is shown in Curve A. Here we are discussing rotational energies; we are ignoring any kinetic energy in the system resulting from the shift of the center of rotation.

Curve C shows that energy is fed into the club slowly at the beginning of the downswing. But as the arms begin to lose energy, the club begins to gain energy at an ever increasing rate. It is obvious that energy is fed from the arms into the club as the energy of the arms decreases.

The Efficiency of the Swing

It may be instructive to describe quantitatively the energy changes in a golf swing, all the while remembering that we are looking at the results of calculations using our model. The overall efficiency of the swing, the ratio of the kinetic energy of the ball as it leaves the club face to the kinetic energy of the arms and club at impact, is estimated for the standard swing to be nearly 30%.

Curves similar to those shown for the standard swing in Fig. 5.1 were examined for the modified standard swings previously discussed. The effect of the modifications can be interpreted in terms of their effect on the efficiencies of the swings. As an example, consider the swing with a wrist-cock angle of 90 deg at the beginning of the downswing. The energy curves look very similar to those for the standard swing, but the efficiency of this swing is less, with an estimated value of 26%. Increasing or decreasing the torque TS applied by the golfer in the standard swing hardly changes the efficiency at all.

This analysis indicates that to obtain the maximum clubhead speed within his muscular capabilities, the golfer must try to find a swing with the maximum efficiency. The adverse modifications discussed in Chapter 4, such as the early uncocking of the wrists, the use of a loose grip, and the early straightening of a bent left elbow, may be considered as lowering the efficiency of the swing.

Chapter 6

Producing Power with Precision

The Challenge of the Course

Most golf courses are designed to challenge the golfer. Usually, the distance between the tee and the cup is chosen such that each shot must be well hit if the golfer hopes to par the hole. There is, of course, much more to golf than just the swing of a club. The three- and four-putt greens, the bunker shot that became necessary through misjudging the effect of the wind, the foolish idea that you can hit from deep rough with enough precision to thread the ball through the small opening between two trees, the errors in the choice of clubs, these and the intangible effects of your particular mental attitude of the moment make the game of golf something a great deal more than just swinging the club. Everyone knows that you cannot play a satisfactory round of golf without being able to hit the ball hard and with precision. How far must a golfer be able to hit the ball in order to hope to play par golf? Where is he to find the needed energy?

No two golf courses are the same, but most courses have the common characteristic of four par-five holes, four par-three holes, and ten-par four holes. As we look for energy for the swing, it is of interest to consider how the distances a golfer achieves with his driver and fairway woods affect his possible score on a hypothetical average course. It will be assumed that the golfer plays perfect golf with two putts on each green and that the distances previously mentioned are the only varying factors. I have used yardages of regulation courses with which I am familiar; if I had used yardages of other courses things might have come out slightly different, but the general conclusions would be similar.

If you are able to average only 160 yards for your drives and wood shots, you will lose 15 strokes to par just because of this lack of distance. A ten-yard increase in the average distance you hit the ball with these clubs would decrease your loss to par to 12 strokes. Table 6.1 shows how the loss of strokes to par gradually decreases for successive ten-yard increases in this average distance. While few play perfect golf, this table shows how important distance with precision can be in improving a golfer's score.

The distance a golf ball travels depends on many factors, but one

TABLE 6.1.

Yards	160	170	180	190	200	210	220	230
Stroke lost to par	15	12	9	7	5	3	1	0

necessary condition for distance is that the ball have a high speed just after it leaves the clubhead. The conservation of momentum principle tells us that for a given club and a given ball, the speed of the ball depends directly on the speed of the clubhead. We have seen that the large muscles of the body must come into play to achieve the tremendous clubhead speed, 100 miles per hour or more, needed to hit the ball far enough to approach the possibility of shooting par golf. The problem, then, is to put the necessary energy into the arms and hands and ultimately into the clubhead by the time the ball is hit and to do this consistently with precision.

Throwing a Discus

This problem has been studied for a long time. Some 3000 years ago the Greek discus throwers knew that they needed to get the hand holding the discus moving at high speed before letting the discus fly. The modern method of throwing a discus is a complicated matter, but I felt that discus throwers might have something important to say to golfers.

A regulation discus is now a flat disk having a mass of 2 kg (therefore it weighs about 4.4 lbs here on Earth) and is 8.625 inches in diameter. It is thrown from inside an 8 ft 2.5 in circle. The thrower starts his throw with his back toward the target and completes one and one-half turns before he lets the discus fly. The goal is to throw the discus as far as possible. He has wide latitude as far as direction is concerned. We are not interested in the details of this turning. Rather, we find that some of the first motions of the discus thrower are of interest because he has found how to supply energy to the discus through the use of the large muscles of the body, those of his legs, thighs, and back. His turning allows him to do this for a longer time than if he were to throw without moving his feet. The golfer's motion is restricted, since he performs his action with his feet essentially at rest so that he may have precise control of the direction of the flight of the ball.

The discus is held in the right hand with a straight right arm and is swung from right to left in the throwing motion. The golf club is held in two hands but is controlled mainly by the left hand and is swung from right to left with a straight left arm. Though these appear to be entirely different motions, they are really surprisingly similar.

In a description of the modern technique of discus throwing we are told that the thrower's feet should be comfortably placed slightly more than hip width apart [16]. From this position, with knees slightly bent, he takes two or three preliminary swings in which his weight shifts easily from one foot to the other. On the last swing to the right before the throw, the discus is swung as far back as possible, bringing the discus thrower into a coiled-up position, with most of his weight supported comfortably on a straight right leg with the left leg flexed slightly at the knee and the left heel slightly raised. In this coiled-up position the shoulders have rotated with respect to the hips, and the hips have rotated with respect to the feet. The thrower's back is straight and more or less vertical.

The reader will recognize the similarity of the discus thrower's position at the start of his throw to that of the golfer's position at the top of his backswing. An essential difference is that the plane of the swing of the golfer is closer to the vertical and that of the discus thrower is closer to the horizontal. However, in each case this coiled-up position allows for a powerful action of the large muscles of the body at the start of the motion. The discus thrower is advised to start the motion by moving his hips ahead of the shoulders and the shoulders ahead of the arm holding the discus. This advice is given undoubtedly to emphasize that the discus throwing is not done by the shoulders alone. One finds similar advice given to the golfer, probably for the same reason.

Getting the Feel of It

I do not think that one person can tell another how to move the various parts of the body to produce a powerful golf stroke. I do think, though, that a person can get a feeling of what he needs to do by trying to give something other than a golf club a high speed in a given direction. Having a club in hand and trying to hit a ball keeps the motion of the club foremost in the mind of the golfer. The purpose before us is rather to give some object a high speed to get the feel of using the large muscles of the body. I recommend taking some object with enough mass to be noticeable and swinging this object, using the left hand, along a curve similar to that along which your hands go in the swing of a golf club. I have used a half-brick; later, I made myself a sand-filled canvas dumbbell to reduce the hazard involved. In this throwing motion the left arm may be kept straight. The motion should be similar to that of the beginning of the swing of the discus thrower. The plane of the swing will of necessity be more vertical than the plane of the first motion of the hands of the discus thrower, and for that reason the position of the hand at the top of the backswing will be higher. But the feeling of coiled tension and the feeling of the large muscles working throughout the downswing will be much the same. The golfer should remember to keep his right leg comfortably straight on the backswing.

He should not sway. With these points in mind the golfer should throw the object directly toward the target, giving it as much energy as possible.

By throwing the object several times the golfer will find what he must do to impart the greatest amount of energy to it. He will soon realize that without his thinking about it, the large muscles of the legs, hips, and back have produced a shift of his weight to his left leg. It has long been recommended that a golfer use such a shift in his down-swing. But here we see that the shift is produced as a result of the use of the large muscles: The shift is a result rather than a cause. With this body motion in mind, the golfer should take a club in hand and repeat this motion making his left hand, along with his right hand, move in the same manner as before. At this point the golfer should not think at all of what the club will do. The attention should be directed entirely on what the body is doing to move the hands. The golfer may be surprised to find that even when the club is completely ignored it will perform automatically to whip around as it does in a good golf stroke.

The Pull of the Left Shoulder on the Left Arm

In the motion just described, the golfer should consider in detail the sensations he has in throwing the object. If the object has enough mass, he will feel in his hand the horizontal pull he must exert to accelerate the object during the swing. The more quickly he swings, the larger the pull will be. He will also feel the pull his shoulders exert on the upper end of his arm. He may feel the tension at his elbow between the upper arm and the lower arm. In swinging a massive object these forces are felt without difficulty. When the massive object is replaced by a golf club, for a similar swing these forces are much smaller. However, the golfer should at least feel the horizontal force of his left shoulder on the upper end of the left arm. This is the force that gives his arm its hori-zontal acceleration during the swing and, if the arm is high in the backswing, also gives the arm an angular acceleration in the down-swing. Some of the work done by this force finally appears as energy in the club.

This horizontal force will be large enough to be conspicuous only if the muscles of the legs, thighs, and back are working during the swing. If the golfer is swinging the club by applying a torque with his shoulder muscles, the force between the shoulder and the arms will only be the centripetal force of the swing, and this force is quite small while the arm is above the horizontal position. The presence of this feeling of a horizontal force of the shoulder pulling on the upper arm is a sure test of whether the large muscles of the body are supplying energy in the stroke.

The feeling of the pull of the left shoulder on the left arm has been emphasized because many golfers are not conscious of this means of supplying energy to the swing. Reference to this aspect of the swing is conspicuously absent from most golf literature. Bobby Jones [8] is the only one I know who even mentions it. I do not intend to imply by this emphasis of the shoulder pull that the source of energy in the shoulder muscles should be neglected. They supply a major portion of the energy to the stroke, but a really powerful stroke cannot be produced by the shoulders alone.

Unlike the golfer, the discus thrower is not primarily concerned with the direction involved in his actions. He has considerable latitude as to where the discus goes. Here we are concerned with the similarities of the motions because we wish to experience the use of the large muscles of the body. Later we shall examine various aspects of the golf stroke to see how the necessary greater precision so important to the golfer may be obtained.

We should now be ready, with our present understanding, to consider how a golfer may put together a golf stroke limited only by his capabilities and not by his lack of knowing what he should be trying to accomplish.

CHAPTER 7

Developing Your Own Golf Stroke

The Golf Swing Is Subject to the Laws of Physics

After a round of golf in my church league, I suggested to my opponent of the day that I would like to coach him for a few minutes behind the clubhouse. He accepted. I showed him what would happen to his nine-iron shots if at the beginning of his downswing he would pull on his left arm with his left shoulder in the direction he wanted the ball to go. He was soon hitting beautiful shots. Some weeks later, having caught up with his foursome on the tee because of a lost-ball delay, I watched him produce a beautiful, yes tremendous, drive, far and in the middle of the fairway. I asked him to supply me with the name of his driving coach. He replied, "I am using the same swing for driving that you showed me to use on my nine-iron shots." The simple suggestion I gave him developed out of my study of the physics of the golf swing. He had made the significant discovery that if you can swing one club, you can swing them all.

The discussion presented so far has mostly involved theoretical aspects of the swing of a golf club. The question remains whether the general understanding coming from applying the laws of dynamics to the swing of a club can be used in the design of a practical swing, one a golfer can use to make golf a more enjoyable game.

It is doubtful that anyone can be told in complete detail, item by item, what must be done to produce a serviceable golf swing. A person can learn only by doing. But this does not mean that the doing should be done in a haphazard fashion. The theoretical understanding of the golf swing puts limits on what should be attempted in the doing, but it does not specify exactly what should be done. If the theoretical understanding did establish exactly what should be done, we should all be swinging like automatons. The theoretical understanding indicates the techniques that should be used, but within the general area of the correct technique, we may each develop our own style of swing. The differences expressed so dogmatically in the various books of golf instruction undoubtedly concern differences in style rather than differences in fundamental technique. The suggestions offered here should help the golfer to explore variations in style within the general area of the correct technique. The correct technique is essentially determined

51

by the laws of physics and the limitations of the human body. Let us face the fact that our bodies are all different, and for this reason alone the style of the stroke that each of us may develop will differ from the style of other golfers. It appears that a great deal of golf instruction is of the nature of teaching a particular style rather than helping a golfer to develop a technically correct style of his own.

Learning to handle a golf club properly is very much like learning to play a musical instrument or learning to ride a bicycle. You may be able to say what you are supposed to do but be completely incapable of doing it. The skill of an activity of this kind comes only after some change in a person's nervous system. Some say that a skill develops only as a learned movement is stored as muscle memory. It is doubtful that anyone knows exactly what happens to us when we learn something, either a manipulative skill or otherwise. We develop a skill by getting the "feel for it." The most rapid approach to skill in golf is very consciously to develop a feeling for things that should be done. We shall look at ways of doing this.

Putting the Swing Together

My intention is to have you experience and feel the swing of a golf club as a dynamic event. Here, as well as on the practice tee, ideas have to be communicated by using words, and this implies that we have to cut the swing up into parts in order to discuss it properly. Dismembering the swing is a very artificial but probably necessary process. The real problem comes when the learner tries to put the various parts of the swing together into a dynamic whole. It appears that the best way to do this is to develop the feeling for the individual parts of the swing and to rely on remembering these feelings as the complete swing is being assembled.

The Grip

At the beginning of a golf stroke a golfer first grasps the club in his two hands in what is called the "grip." Most golf instruction starts with an elaborate discussion of the grip. The grip is indeed an important aspect of the golf stroke, but we shall postpone discussing it until later. Some expert golfers hold the club one way, others hold it another way. If you wish to follow along with a club in your hands, you may use your present grip without worrying about what your own grip will ultimately become. I suggest that you should first develop your swing, the one that feels right for you and is repeatable, powerful, and precise, and then find how you must grasp the club so that as you come through the ball, the face of the club is square to the intended path of the ball. Starting with a grip that someone else has found to be suitable for his swing may constrain your movements and interfere with the development of your swing. When you find the grip you intend to use for normal strokes, you may, of course, modify it for intentional hooks and slices.

The Stance

Let us start from the ground up and look at the many things that have to be considered in the development of a good golf stroke. They cannot all be kept in mind at one time. They have to be thought about one at a time, practiced individually so that the feeling is established, and as progress is made, checked and rechecked so that they become almost automatic.

The way the feet are placed on the ground is called the "stance." If the toes are even with the line that goes through the target, the stance is called "square." If the right foot is drawn back a bit, the stance is called "closed." If the left foot is drawn back a bit, the stance is called "open." A golfer should probably use a square stance unless he finds a good reason for some slight modification. If a golfer develops a good swing, he does not need to adjust his stance. If his swing is questionable, he may be advised by some to modify his stance in an attempt to rectify some difficulty. Would it not be better to attempt to correct the fundamental error first?

When a golfer addresses the ball with his driver in hand, his feet should be placed so that the distance between the insides of his heels is no more than the width of his shoulders. He may take a stance that is much too wide, apparently with the thought that a wide stance will offer a much better base for moving the hips. Actually an excessively wide stance will hinder the proper motion. A quick measurement with a yardstick may show you that your own shoulders are not as wide as you think they are. The toes may be pointed in the direction usually assumed in walking or standing, although many golfers prefer to have the right foot placed at right angles to the line toward the target and the left foot turned somewhat toward the target.

How the Golfer Should Stand

After the golfer has placed his feet properly, he should squat slightly by flexing his knees, and he should bend forward at his hips. The question is how much there should be of each. Some golfers stand very upright to the ball, and others seem to exaggerate the forward bend of the hips.

If the angle of forward bend is measured from the vertical to the line drawn from hip joints to the midpoint of the neck, this angle is found in photographs in the literature to vary among expert golfers from about 25 deg to more than 40 deg when they are using the driver. The forward bend at the hips is thus a matter of style, and a golfer should choose an angle that feels right to him. When using woods he probably should experiment by bending forward from 30 to 35 deg and explore variations from this position. He may wish to use photography in this study. He will find that he needs to bend forward farther when using the shorter clubs.

The flexing of the knees should be held to the least amount that will allow the golfer to rotate his hips freely. He will find that he cannot turn properly if he stands stiff-legged to the ball. However, if he exaggerates the flexing of the knees, he may be bothered by a reflexive action during the swing of the club tending to straighten his legs. When a well-swung driver passes through the bottom of the swing, the centrifugal force produced may make the club and arms pull on the shoulders with a force up to and perhaps greater than 100 lb; this force has to be supported by the golfer's legs. Either through a direct reflexive action, the kind we have all experienced in the doctor's office in the well-known knee-jerk reflex or through the uncontrollable anticipation of needing to support that extra force, the legs tend to straighten as the clubhead comes up to the point of impact with the ball. If the knees are bent too much, the straightening of the legs will raise the clubhead and cause it to move on a higher path than was intended. Keeping the flexing of the knees to a minimum consistent with the free turning of the hips will keep the lifting of the clubhead to a small amount even if the reflexive action does occur. There are photographs of expert golfers showing this action exaggerated to the extent that they are up on their toes at the bottom of their swings [17]. In the interest of precision such aberrations of style should be avoided.

The Position of the Hands

After he has taken the position just described, the golfer must decide how far his hands should be from his body. When he addresses the ball with the driver in hand, his hands should be placed so that the thumbnail of his right hand is very near the vertical plane containing the target and the golfer's eyes. Some expert golfers hold their hands a little closer to the body than this, and some hold them a little farther away. The placing of the hands is a matter of style, but variations of style should not include positions in which the golfer feels that he is reaching for the ball. For shorter clubs the hands are placed somewhat closer to the body.

The Distribution of Weight

The question that naturally arises next in the mind of the golfer is how his weight should be distributed while addressing the ball. In the literature on golf you will find many answers to this question. Since the centrifugal force on the golfer's shoulders will pull him forward some-what toward the bottom of his swing, he should stand with about equal weight on his two feet with the weight slightly more on his heels than on his toes. The diagram in Fig. 7.1 shows how the feet may be placed and how the weight may be distributed. The size of the black circles indicates the relative amounts of the weight on the heels and toes. Such a weight distribution gives the golfer a feeling of stability at the start of the backswing.

The Position of the Ball

The next question confronting a golfer is where to place the feet in relation to the ball. For the swing with a driver, everyone agrees that the ball should be placed opposite the inside of the heel of the left foot. For other clubs there is no such agreement. Some think that as we move toward the shorter clubs the ball should be moved successively to the right relative to the feet; others think that the ball should always be placed in the same position opposite the left heel for all clubs. If a

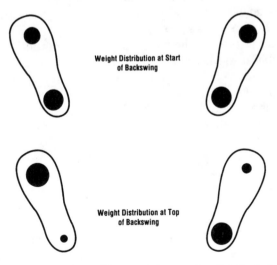

FIGURE 7.1. (Upper) how the golfer's weight may be distributed at the start of the backswing. (Lower) how the weight may be distributed at the top of the backswing.

golfer adopts this constant position of the ball, he will always address the ball in the same way, a procedure that can be done with considerable precision. If he varies the position of the ball from club to club and opens his stance, as he must to compensate for the fact that the ball will be hit earlier in the swing, before the clubhead would otherwise be traveling toward the target, he may find it difficult to prepare the swing with the same precision. The fact that both procedures are used by competent golfers indicates that placing of the ball for the shorter clubs is a matter of style.

The Lateral Position of the Hands

After the golfer has placed the head of the club behind the ball and has taken his stance, he must decide on the lateral position of his hands. Some golfers place their hands so that the club shaft points toward their spinal column; others place their hands so that the club shaft and the left arm appear to be in a straight line to the left shoulder. Some place their hands in an intermediate position. If one is to take advantage of what the calculations have shown and what many golfers have found from experience in regard to having the hands ahead of the ball at impact, it seems reasonable to place the hands ahead of the ball at the start of the backswing and grip the club while the hands are in that position. Such a position should ensure that the club face is square to the intended line of flight of the ball when the clubhead returns to the point of impact. The placing of the hands at the start of the backswing is again a matter of style, and each golfer must decide from his own experience what he wishes to do.

As the golfer addresses the ball, his left arm should be comfortably straight, and because his right hand grips the club below the left hand, his right shoulder will be slightly lower than his left shoulder. The golfer should find himself addressing the ball without any undue tensions. He should not be gripping the club more tightly than necessary to keep it from slipping from his hands. The muscles in his arms and legs should be able to move freely.

We are now knowledgeable enough to watch with a critical eye how a golfer addresses the ball. By watching him we cannot tell exactly how he distributes his weight or how tightly he grips the club, but we can judge how his positions fall within the generally acceptable areas. The reader may likewise stand up to the ball and assume similar positions, but now he is able to feel, among other things, the distribution of weight, the tightness of the grip, and the muscular tension in the flexing of his legs. As the reader gradually develops his own swing, he should be able to recall these feelings so that his swing can be repeated with precision.

After the golfer has addressed the ball in his own particular style, he will probably waggle the club a bit and then go directly into the swinging of the club. This waggle we shall consider later. The first part of the swing, the backswing, goes smoothly into the second part of the swing, the downswing. This separation into two parts for discussion should not imply that the golfer swings the club to the top of his swing as one distinct dynamic event, and then swings the club down to hit the ball as another. After the waggle is completed and the swing is started, each part of the swing is smoothly connected to each succeeding part.

The Backswing

At the start of the backswing, the hands and arms move as one body. The clubhead is swung on a plane tilted somewhat from the vertical. The angle of tilt is a matter of style. Some golfers have a more upright swing; some have a flatter swing. Shorter clubs are usually swung in a more vertical plane. The plane in which the clubhead is swung is determined by the relative amounts of turning of the hips and turning of the shoulders. If the hips alone are rotated about a vertical axis, the clubhead will move in a very flat swing, whereas if the shoulders swing the club up with the hips fixed, the clubhead will move in a very upright swing.

The clubhead should be started back with a combination of hip turning and shoulder turning. In turning the hips, the left leg is bent somewhat, bringing the left knee forward. This brings the weight of the left leg forward onto the left toes. At this point in the swing, some golfers have a tendency to move the body toward the ball and also to move the weight on the right foot onto the right toes. This forward movement of the body should be prevented. This can be done by making sure that the weight on the right foot is shifted a little more onto the right heel. This shift in weight forward onto the left toes and backward onto the right heel is illustrated in Fig. 7.1. This shift in weight will keep the center of gravity of the body from moving as the hips are turned. Many golfers keep the heel on or close to the ground. Probably, a pronounced lift of the left heel should be considered an unacceptable aberration of style.

The literature of golf instruction abounds with discussions of how during the backswing the golfer's weight should be transferred onto the right foot in order that during the downswing the weight may be shifted back onto the left foot. At the same time, the golfer is admonished not to sway on the backswing. These two statements appear to be incompatible. If weight is transferred from the left foot to the right, this means that the golfer's center of gravity must shift, too. The body's center of gravity must move to the right, and this is the sway we must avoid. From the physics of the situation we may conclude that this supposed transfer of weight to the right foot is really simply a shift of weight already on the right foot further back onto that heel. The shift of the golfer's weight illustrated in Fig. 7.1 is conducive to a golf stroke in which the golfer can maintain his balance and swing the club on the backswing without swaying.

As the backswing progresses the left arm is kept straight and the right arm will fold toward the body with the elbow pointing down and back. When this happens, the wrist will begin to cock naturally, and the momentum of the clubhead will keep on cocking the wrists until the backswing is completed. The limits of the backswing will be determined by the flexibility of the golfer's body. The limits of the wrist cock will be determined by the flexibility of the golfer's wrists in conjunction with the momentum of the clubhead as determined by the quickness of the backswing. An overly slow backswing will afford little momentum to help cock the wrists. As he approaches the top of the backswing, the golfer should avoid two common faults, the bending of the left arm and the loosening of the grip of either hand. He should reach the top of his backswing having kept his center of rotation at rest and with his right leg still slightly flexed.

The Downswing

We have now reached the crucial part of the golf stroke, the start of the downswing. Everything the golfer has done so far has been in preparation for turning on the power with the necessary precision. How is this to be done? It does not just happen.

If you look in the golf literature for advice on how to proceed with the downswing, you will find many suggestions, all of which appear to be misleading. Here is a list of some of them for our consideration.

1. Pull down on the shaft of the club as though you were pulling on a rope leading to a bell in the church belfry.
2. Pretend you are going to push the upper end of the shaft of the club into the ball on the tee.
3. Swing the clubhead with your hands.
4. Rotate the body by raising your left shoulder and lowering your right hip.
5. Push off with your right foot.
6. Start with a lateral shift of the hips to the left.
7. Shift quickly over onto your left leg.
8. Turn your hips out of the way of your hands.
9. Start the downswing by rotating your hips to the left.

The first three bits of advice lead to the production of a relatively powerless stroke because the shoulder muscles become the principal source of power. The other bits of advice on the list, individually or in combination, will not lead to a powerful, precise stroke because the suggested motions can be made with hardly any club motion at all. However, when a golfer has developed a powerful stroke, he will appear to some extent to have all the motions suggested. In my opinion these motions come along with the powerful, precise swing, and not a single one can be considered as describing the essential nature of the swing.

How, then, are we to describe the motion leading to the optimal downswing? May I suggest that the downswing should be organized along the lines developed in our second look at the golf stroke. The large muscles of the legs, thighs, and back supply energy to the left arm at least at the start of the downswing. The shoulder muscles start their action gradually at the same time and continue to supply energy to the left arm throughout the rest of the downswing. Later, the large muscles slow up the motion of the body, thereby giving an additional amount of energy to the system. The dynamics of the swing will transfer energy from the arm into the club; the golfer does not have to worry about this aspect of the swing at all.

The downswing of any golfer may be described as a combination of three separate motions. One of these motions is a rotation of the arms about an axis perpendicular to the plane of the swing. A second motion is a rotation of the body about a vertical axis. A third motion is the shift of the body toward the target.

Let us consider strokes in which each of these motions is dominant. In the swing where the emphasis is mainly on the use of the shoulder muscles to give the arms a torque about the plane of the swing, at the bottom of the downswing the golfer's arms, hands, and club will be moving toward the target with considerable momentum. The golfer's body will recoil in the direction away from the target with the same momentum unless he braces himself to prevent this recoil. The physics of this process is exactly the same as in the firing of a gun. The gun recoils with exactly the same amount of momentum as the bullet but in the opposite direction. When the golfer is not properly braced in anticipation of the recoil there will be a "falling away from the shot." Watch your fellow golfers and you will observe this type of swing much too often. Such a swing does not produce a high clubhead speed and a well-hit ball.

The amount of force the golfer exerts horizontally on the ground determines how much recoil he will experience. He may not fall away from the shot in a conspicuous manner, but he will likely not end the swing with his weight mainly on his left leg.

In the downswing in which the rotation about the vertical axis is the dominant motion, the direction of the clubhead is likely to be uncertain. For the perfect hit, the clubhead must be moving to the ball along the line to the target. When the body is rotating in this way, it is difficult to make the clubhead meet the ball while it is moving in the proper direction.

In the downswing in which the shift of the golfer toward the target is adequate, the shift motion will help to produce a relatively straight path of the clubhead through the lower part of the swing, and if that motion happens to be toward the target, the ball will go in the proper direction.

We have arrived at the point where we can begin to answer the question of how the golfer can produce a downswing with the necessary power and precision. He must find the proper combination of these three motions that will allow him to bring the clubhead to the ball along a line accurately directed toward the target. Later, we shall show that the golf stroke that will send the ball directly toward the target cannot have the clubhead moving "from the inside out" or "from the outside in" on impact with the ball. Precision in the path of the clubhead is reflected in the precision in the path of the ball.

How to Tell Whether the Path of the Club Is Correct

How is the aspiring golfer to know whether the clubhead is moving in the proper direction in any particular downswing? There are two ways. The flight of the ball is determined, except for the effect of the wind, by the collision between the clubhead and the ball. We shall discuss this topic in some detail in Chapter 9. But when the golfer understands the cause of hooks and slices, pulls and pushes, he can watch the flight of the ball and thereby infer what the clubhead was doing at the crucial moment of impact. He can also tell what the clubhead is doing at impact by watching it in its motion. The clubhead moves so fast that it is difficult to see it unless means are taken to make the clubhead visible. If a piece of white tape is put across the top of a wooden clubhead, parallel to the path of the club, and the club is swung above a dark background such as a rubber mat or a black cloth, there is enough afterimage to allow a person to see the direction the clubhead is moving at the bottom of the downswing. A club decorated in this manner would likely not be a legal instrument in tournament play, but I have found such a club valuable help in examining my own swing.

It may be of some value for a golfer to try the three kinds of swings previously discussed so that he can get a feel for them. In arriving at a proper downswing, the golfer must combine these motions in the proper proportions and in the proper sequence. No one can tell him, except in general terms, how he is to do this. He can best proceed by experimenting.

Thoughtful Practice

The golfer who is earnestly trying to improve his swing undoubtedly brings to the practice tee many well-established habits that make any change in his swing most difficult. He has many things to think about. The backswing can be done slowly enough, perhaps, that he can think through what must be done to arrive at the top of his swing in the proper fashion, but the downswing happens so fast that possibly only one of many aspects of the downswing can be kept in mind during any particular swing. This means that progress toward a really good repeatable downswing will come only after much thoughtful practice. The practice must be done with understanding and with the intention of establishing new habits. As habits are formed, the golfer will find that there are fewer and fewer aspects of his stroke that he must keep at a high conscious level.

In view of the fact that many golfers do not use the large muscles of the body as they should be used, a golfer should start the work of improving his stroke by keeping uppermost in his consciousness the feeling of the horizontal pull of the left shoulder on his straight left arm during the start of the downswing. A strong smooth pull by the shoulder in the direction of the target will ensure a powerful stroke with a high finish of the hands and a proper shift of weight to the left leg.

The golfer will find that the clubhead very nearly follows a path that is parallel to that taken by the hands. If the swing is designed so that the hands come through their lowest arc moving toward the target and continue to move along this path as long as possible, the clubhead will automatically follow the correct path, and the golfer will have the high finish so necessary in a precise stroke. If the golfer develops the habit of watching the clubhead go through the lowest point of the swing, he will reap a valuable bonus.

Start from a Quiet Center

The consequence of simply turning at the beginning of the downstroke to look to see where the ball is going to go points up the utmost importance of keeping a quiet center of rotation at least until the downswing is under way. A golfer can prevent swaying by keeping his head at rest during the backswing. He may find that swinging the club back without moving his head constrains him so much that he feels he cannot put much power into the stroke. He may feel that moving his whole body to the right, his head included, puts him well behind the ball and in a position where he has a feeling of almost unlimited power. This feeling comes from his anticipation of swinging his left arm using his shoulder muscles. But if he is to develop a precise and powerful stroke, he must learn to live with the feeling that comes when he keeps his head quiet during the backswing.

Check Head Motion

In developing your own swing, how are you to tell whether you swing the club back with a quiet head? Some advised having a friend stand in front of you and hold his hand on your head while you practice your stroke. Your friend can tell you whether you are swaying or not. I have noticed that swaying is a very prevalent habit and needs to be checked frequently. There is a very simple way to check head motion. Simply stick a pencil vertically into the ground a few inches beyond the ball and lay a wooden tee on the ground a few inches beyond the pencil placed so that as you address the ball you may look over the top of the pencil and see the point of the tee. In other words, your eye, the top of the pencil, and the point of the tee are in line. Any motion of your head will cause an apparent relative motion of the pencil and the tee. If this relative motion, called parallax, occurs, you are swaying. This use of parallax is a very sensitive test for swaying.

During the downswing, if the large muscles are to be used to supply energy, the center of rotation of the swing must move to the left. Photographic studies of expert golfers such as that in Fig. 2.1 indeed show that a shift of the center of rotation does take place early in the downswing. The golfer should not feel that he is rigidly constrained during the downswing, but neither should he feel that he can move the center of rotation with complete abandon. The limits of possible motion should ultimately be decided by what happens to the ball.

Starting the Swing

After the golfer has taken his stance, he faces the problem of getting the swing under way. The discus thrower is advised to take two or three practice swings before the final one. Practice swings, after the golfer has set himself up to the ball, would also be of value. But of course, the ball is in the way. The practice swings of the discus thrower allow him to recall the muscle memory of previous throws and to ensure a correct and comfortable position from which to start the throw. Golfers have substituted the waggle for the same purpose. The waggle is not an exhibition of jitters, although to the nongolfer it may appear so. The waggle is a means of bringing up muscle memories of previous similar swings and of checking the position that has been assumed for the swing to come. The exact motions in the waggle vary greatly from golfer to golfer, and for a particular individual from stroke to stroke. Whatever the motion, the waggle seems to be a necessary part of the swing, for few golfers just stand up to the ball and hit it. A golfer must devise his own individual waggle, the one that serves him best.

Three Styles of Grip

Now we come to a consideration of the grip, the way the golfer makes connection with the club. There is considerable misapprehension caused by the usual discussion of the grip. One is led to think that the misplacing of a single finger by the least amount is a matter of life and death. Some excellent golfers use one grip, some another. A professional golfer when asked about the use of the ten-finger grip, where all ten fingers are on the club, replied that some golfers on the tour do use this grip. He said he had used it himself for one season and had concluded that for him it did not give quite the precision he wanted, although he could get more distance with it. He said that he thought the ten-finger grip would be best for most casual golfers.

There are three styles of grips. In one, the fourth (little) finger of the right hand is placed between the first and second fingers of the left hand, and the first finger of the left hand is placed between the third and fourth fingers of the right hand. This interlocking grip is very compact. In another, the fourth finger of the right hand rests outside the left fist in the depression between the first and second fingers of the left hand. Most golfers grip the club this way. In the ten-finger grip, all the fingers are in contact with the club. In all of these grips the club runs diagonally across the left hand and is gripped by the fingers of the right hand. The left hand grips the club so that the back of the hand at address is toward the target, and the right hand grips the club with the wrist neither rotated to the right nor to the left. The exact orientation of the hands will depend on how far ahead of the ball the hands are at address.

When the golfer grips the club this way with any of the three styles mentioned, he should find that as the hands reach their highest point on the backswing, the club is held so that the cocking of the wrists produces no great strain on them. Two criteria of a proper grip are that it allows a full wrist cock and that it allows the clubhead to come in contact with the ball with the club face square to the intended line of flight. A person learning to swing a golf club should be encouraged to experiment with various grips in the process of finding one suitable for him.

Develop Your Swing on the Basis of Feel

We are now knowledgeable enough to watch a golfer complete his golf stroke. Here the action is so rapid that we are able to notice perhaps only one characteristic for each swing. If you watch his hands to see something of how he handles his wrist cock, you will be unable to tell what he does with his left heel. Even after observing swing after swing, you will not be able to form much of an idea of what he "feels." If you are to develop your own swing on the basis of how it feels, you must do it yourself. Of course, it helps to have someone point out your errors and put you back on the right track on occasion. But you are the only one who can put your stroke together.

As a golfer makes progress in developing his stroke by establishing new habits, he will find fewer and fewer things about the stroke he needs to keep in mind. But relapses will occur. If he has developed a checklist, perhaps he will be able to determine which of the old habits have returned. He may find it best to turn to a knowledgeable instructor for help.

The reader should realize that I have not presumed to tell him exactly how he should swing a golf club. I doubt that anyone can tell exactly how it is to be done. My intention, rather, has been to outline, as suggested by the application of physical principles, the various characteristics of a good swing and indicate the variations of style that may be considered to fall within acceptable limits. Mastery comes only as the feeling of the swing is crystallized through much practice. Practice brings dividends, according to Bobby Jones, only when the golfers "start out with an accurate conception of what they want to do." I hope that my readers have been helped along the way toward acquiring such a conception.

The principles of physics may be applied to other aspects of golf. The action of the club on the ball and the action of the air on the ball in flight may be analyzed to bring some understanding of what the ball does while it is in contact with the club and why it acts as it does while it is airborne.

CHAPTER 8

The Aerodynamics of Golf

Aerodynamic Forces

The use of aerodynamic forces by the expert golfer was demonstrated to me once by a drive off the tee by a well-known professional, Arnold Palmer. I stood directly behind him so that I could have the best view of the flight of the ball. Instead of the ball going on the parabolic path I had expected, it climbed in a nearly straight line for about 3 seconds and then began to fall. Recall that Newton's first law states that a body continues in a straight line at constant speed unless a force acts on it. The nearly straight path indicated an almost zero net force acting on the ball in the vertical direction. Where did this lifting force about equal to the weight of the ball come from? It was the aerodynamic force on the dimpled spinning ball, traveling at a high speed, that was balancing the vertical force of gravity.

Anyone who has ever hit a golf ball with a club realizes that the flight of the ball is affected by these aerodynamic forces. Without the force of the air on the ball, the ball would travel on a very different path than it actually does. If you toss a ball a short distance, it will move accurately on a parabolic path because the force exerted by the air on a slowly moving ball is small. When a well-hit ball travels at speeds up to 140 miles per hour, the force of the air on the ball is not small. In some cases this force can become even larger than the weight of the ball and can produce spectacular modifications in its flight path.

The Early Golf Ball

The early golf ball was a spherical leather pouch filled, while wet, with wet goose feathers [1]. The pouch was stitched with linen thread, turned inside out so that the stitching was on the inside, and filled with all the feathers that could possibly be forced into it through a small hole, which was then carefully closed by a few more stitches. The ball dried to become very hard. It was then oiled and whitened to become an expensive golf ball. The ballmaker could make only four or five in a day. An acceptable drive with such a ball was from 150 to 175 yards. These balls became useless when wet.

These "featheries" were standard until about 1848, when the gutta-percha ball arrived on the scene. Gutta-percha is the dried gum of the sapodilla tree. When heated, this material takes on the consistency of putty and can be rolled into a solid ball. These balls were first rolled by hand but later were formed in a metal mold. At first, these balls had smooth surfaces.

The "guttie" balls had two drawbacks; they sometimes broke into pieces, as some modern solid balls have been known to do, and they did not fly as well as the feather balls. As these balls were used, their surfaces became roughened, and in this condition they flew much better than the smoother balls. Soon gutties were being molded with rough surfaces. Experience showed that a rough surface on a ball had a marked effect on its aerodynamic characteristics.

Remove the Dimples and Compare

It is an interesting experiment to remove the dimples on a modern ball using sandpaper, give it several coats of white spray lacquer so that it becomes quite smooth, and then compare its aerodynamic characteristics on the practice tee with a similar ball with its dimples intact. You will be surprised at the much greater distance achieved with the dimpled ball.

The Experiences of P. G. Tait

The first serious thinking about the aerodynamics of golf balls appears to have been done, starting in 1887, by the British scientist P. G. Tait [18]. Professor Tait held a chair of natural philosophy at Edinburgh University, Scotland. Natural philosophy was the name for what is now known as physics. Golf was his favorite recreation.

Professor Tait showed through his studies the importance of spin on the flight of a golf ball. He relates that in his youth he was taught that "all spin is detrimental," and he practiced assiduously to master the art of hitting a ball almost free of spin. After he had completed his researches he wrote, "I understand it now, too late by 35 years at least."

One of his sons, F. G. Tait, a famous amateur golfer, helped his father with the experimental aspect of his researches. Of course, Tait worked before the days of high-speed photography and other sophisticated laboratory apparatus. He devised an ingenious way to make measurements of the spin of a ball. He wrote, "When we fastened one end of a long untwisted tape to the ball and the other to the ground, and induced a good player to drive the ball (perpendicularly to the tape) into a stiff clay face a yard or two off, we find the tape is always twisted; no doubt to different amounts by different players—say from 40 to 120 or so turns per second. The fact is indisputable."

The Lifting Force on a Spinning Ball

Professor Tait states clearly that a ball driven with spin about a horizontal axis with the top of the ball coming toward the golfer has a lifting force on it that keeps the ball in the air much longer than would be possible without spin. He explained the source of hooks and slices in the rotation of the ball about axes other than horizontal.

He wrote, "I have been very, perhaps even unnecessarily, cautious in leading up to this conclusion—I have the vivid recollection of the "warm" reception which my heresies met with—from almost all good players to whom I mentioned them. The general feeling seemed to be one in which incredibility was altogether overpowered by disgust."

Though Professor Tait may have felt that his work was not appreciated by the golfers of his day, his work did open up the field of scientific research on the aerodynamics of golf. As far as I can determine, Tait did not consider the effect of the roughness of the surface of the ball on its motion through the air. Though rubber balls were beginning to come into use about the time of his death, he did all his work with gutta-percha balls.

Roughening the Surface Helps Lifting Force

Shortly after Professor Tait's researches were published, other investigators began to study the effect of the surface roughness on the lift on a spinning ball in flight. The markings on balls went through extensive development before the present dimpled surface was considered to be near the optimum design.

While Professor Tait was apparently the first to study the lift on a spinning ball in the game of golf, others before him had studied this effect. Isaac Newton, some 220 years before Tait, recognized that a tennis ball "struck with an oblique racket" would move on a curved path. This effect is sometimes called the Magnus effect because Heinrich Magnus did some careful experimental work on it. Students of elementary physics study a more general effect discovered by Daniel Bernoulli in 1738. Bernoulli found that whenever the speed of a liquid or a gas is caused to increase, its pressure decreases. This curious, almost paradoxical, effect can be shown to follow from Newtonian mechanics quite simply for the special case of a nonviscous incompressible fluid in streamline flow in a pipe.

Effects of Aerodynamic Forces

Rather than discuss the Bernoulli effect in a theoretical fashion, it is probably better to suggest that those interested try a very simple experiment to demonstrate. Cut a small square piece of light cardboard, about 2 in on a side, stick a pin or a thumbtack through its center, and place it over the end of a spool with the pin in the hole of the spool. Blow through the spool as hard as you are able. Practically everyone expects the card to be blown away, but actually, the harder you blow, the tighter the card clings to the end of the spool. The pin or thumbtack is essential to keep the card from slipping sideways from the spool. The behavior of the card can be understood in terms of Bernoulli's principle. As the air flows between the card and the end of the spool its speed increases, with a consequent lowering of the air pressure in this region. The air pressure on the other side of the card is greater, and there is thus a resultant force from the high-pressure region holding the card firmly against the end of the spool.

There are many other illustrations of Bernoulli's principle. A light ball such as a ping-pong ball may be supported on a small jet of air. Larger rubber balls may be supported on an air jet from a vacuum cleaner used in its blowing mode. In each case the flow of air over the top of the ball produces a low-pressure region. The low pressure resulting from the increase in speed of the air over the top side of the wing of an airplane in conjunction with the high pressure below the wing produces the lift on the wing. An aspirator, whether one to apply medicine to the nose and throat or one in a spraying outfit, uses the low pressure produced in a constricted tube where air is speeded up to lift the fluid being sprayed from the supply vessel to the air stream. As we shall see, the lift on a golf ball is another example of the Bernoulli effect.

Understanding Forces on a Spinning Ball

We shall develop an understanding of the aerodynamics of a golf ball by easy stages. It is a complicated subject and as far as I know is understood only in a qualitative manner. In other words, one does not sit at a desk and with the use of a computer calculate the optimum design of the dimples on a golf ball.

Consider first a ball similar to a golf ball except that it has a smooth surface. Let it be at rest and let it have air moving past it with a small velocity. This is equivalent to having the ball move with the same small velocity through air at rest. However, we shall assume in our discussions that the ball remains at rest and the air moves past the ball, because it appears that this relative motion is easier for most of us to visualize. A possible flow pattern of the air past a ball is shown in Fig. 8.1(a). The lines drawn in the figure show paths of small masses of air in their motion past the ball.

Several figures showing patterns of airflow past a ball are presented in this discussion. These should be looked upon as diagrams suggesting what the flow may be like. Flow patterns past some obstacles have been studied experimentally [19], but as far as I know, no one has thought of how a flow pattern past a high-velocity spinning ball might be observed. These diagrams show patterns of flow that appear to me to be reasonable as a result of my studies and are to be considered only as a basis for discussion.

Streamline and Turbulent Flow

The motion of the air shown in Fig. 8.1(a) is approximately streamline flow. Streamline flow is to be distinguished from turbulent flow. When a candle has been snuffed out, the smoke rises smoothly in still air for a few inches and then breaks into a disturbed motion. The lower part shows streamline flow, while the upper part shows turbulent flow. As the air flows past the ball, the air close to the ball has its speed increased from A to B and then decreased from B to C. This means, according to Bernoulli, that the pressure of the air near the surface of the ball decreases from its value at A to a lower value at B and then rises to have at C the same value it had at A.

The Boundary Layer

For the example now being considered, that for air moving with a very small velocity, the forces of the air pressure on the ball are symmetrical forward and backward and therefore will result in no net force on the ball. But experiments show that there is a force on the ball for small air velocities. This force comes about because air is a viscous fluid. There is a thin layer of air near the surface of the ball, called the boundary layer, in which the speed of the air varies from zero at the surface to the larger value away from the surface out in the streamline flow. The air in

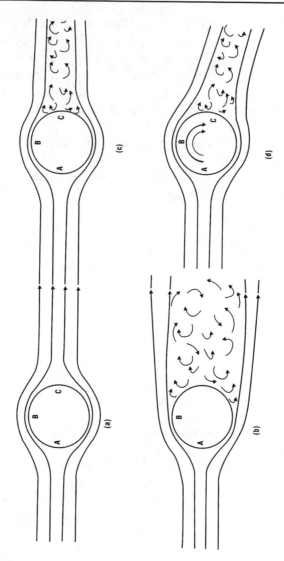

FIGURE 8.1. (a) Possible slow streamline flow of air past a smooth ball; (b) possible flow of air at a higher speed past a smooth ball showing turbulent flow downstream from the ball; (c) possible turbulent flow past a dimpled ball; (d) possible flow downstream from a rotating dimpled ball.

contact with the surface of the ball does not slide over the surface; rather, it is at rest relative to the surface of the ball. Momentum is transferred across this boundary layer from the flowing air to the surface of the ball. The force on the ball is determined by the rate of this momentum transfer.

The Viscous Drag on the Ball

A feeling for what is happening in the boundary layer may be had by imagining the stirring of a kettle of some viscous fluid, such as molasses, with a spoon. As the spoon stirs the fluid, the kettle will turn along with the fluid unless the kettle is held fixed. A torque must be applied to the kettle to keep it from turning. The tangential force of the fluid on the inner surface of the kettle results from the internal viscous effects in the fluid. A similar tangential force on the surface of the ball results from the internal viscous effects in the boundary layer at the surface of the ball. The force on the ball is proportional to the relative velocity of the air past the ball at this low velocity.

In the case under discussion, the air flowing from A to B outside the boundary layer is going from a high-pressure region to a low-pressure region, and we may look upon this pressure difference as helping to increase the air velocity. However, in flowing from B to C, the air moves from a low-pressure region to a high-pressure region and loses velocity in going against this pressure difference. When the viscous effect in the boundary layer becomes large enough that the air near the surface of the ball is stopped before it reaches C, turbulent motion takes the place of streamline flow. This happens sooner or later as the air velocity increases.

The velocity of the air past a well-hit golf ball in flight is much greater than that at which streamline flow will occur. Figure 8.1(b) shows flow lines for small masses of air past a smooth ball without spin when the air velocity is such that extensive turbulence occurs. From A to B the flow is very similar to that in the previous example. At B or a little before B the boundary layer becomes stalled, and a turbulent wake extends downstream from the ball. In this turbulent wake there is considerable violent stirring of the viscous air, and energy is dissipated. When a ball flies through air at rest, this energy dissipated in the turbulent wake comes from the energy of motion (kinetic energy) of the ball. There is thus a resistive force, called "drag," on the moving ball that is not related to the viscous force in the boundary layer but comes rather from the difference in the pressure on the front and back of the ball. The drag in this velocity range varies closely as the square of the relative velocity of the air and the ball.

Dimples Increase Turbulence in the Boundary Layer

Next consider the change in the flow line pattern for small masses of air when the surface of the ball is changed from smooth to the usual dimples found on a golf ball but still with the same large air velocity. The dimpled surface makes the boundary layer turbulent; it stirs the air up a bit. Instead of stalling near B, as in the previous example, the rapidly moving air carries the turbulent boundary layer along with it, helping it to extend further along the surface of the ball from the low-pressure region at B toward the higher-pressure region at C. This is indicated in Fig. 8.1(c).

The turbulent wake starts farther back along the surface of the ball and is thus smaller in cross section than in the case of the smooth ball. The drag on the dimpled ball is considerably smaller than that on the smooth ball; less energy is dissipated in the smaller wake. This has been shown to be true in actual experimental comparisons of the drag on smooth and rough spheres [20].

The Spinning Ball

So far, our discussion has been concerned with the drag on a ball that is not rotating. Professor Tait, as well as all those investigators who came after him, found that a properly hit golf ball acquires spin about a horizontal axis. The ball spins in such a sense that the bottom of the ball appears to be moving away from the golfer while the top of the ball appears to be coming toward the golfer. Sometimes, when a ball is hit above center, it spins the other way. Such a ball is said to have been "topped."

The flow of air past a spinning dimpled ball should be something like that shown in Fig. 8.1(d). The air is assumed to flow from left to right if we think of the ball being at rest. The ball is moving from right to left if we think of it moving through still air.

The turbulent boundary layer is now moving with the surface of the ball as it spins. This means that the air over the top of the ball is moving more rapidly relative to the ball than at the bottom of the ball. According to Bernoulli's principle, the pressure above the ball is less than that directly below the ball. There should thus be a force, called lift, perpendicular to the direction of the ball's motion.

The existence of this pressure difference has been shown to have more than a theoretical basis. Professor J. J. Thomson, the discoverer of the electron, was interested in golf and reported on an experiment, performed at a public lecture, in which he demonstrated a pressure difference in the direction to account for the Bernoulli lift on a spinning ball in flight [21].

As it is pulled along over the top of the ball, the turbulent boundary layer stalls farther down on the back side of the ball, while the boundary layer on the underside of the ball is prevented from remaining next to the ball and stalls even before it reaches the lowest point of the ball. The wake behind the ball thus starts down lower than the wake behind a nonspinning ball. The flow pattern takes on a downward component. The air thus receives some downward momentum, and the ball recoils in the upward direction. This is another way of looking at the origin of the Bernoulli lift on the ball.

Measurements have been made on the aerodynamic lift and drag on spinning golf balls [20,22,26]. It is difficult to give a quantitative description of the results of these experiments because the drag and the lift depend on at least three variables: the speed of the ball through the air, the rate of the spin of the ball, and the surface texture of the ball (smooth and with various dimple patterns). We may, however, make some very general statements.

The drag on a ball increases with its speed through the air and with its rate of spin. It may become about as large as the weight of the ball. The lift on a dimpled ball increases with the rate of spin at a given air speed and increases with air speed at a given rate of spin. It also may become nearly as large as the weight of the ball.

Every golfer knows that a ball hit into the wind will not go as far as one hit with the wind; this is a practical indication of the increase of drag with air speed at the ball. Every golfer knows that a ball hit into the wind will move on a higher trajectory than one hit with the wind; this is a practical indication of increase of lift with air speed.

Competing Effects of Lift and Drag

From what we know about lift and drag on a spinning dimpled ball, it appears that there are two competing effects due to spin on the flight of a ball. A larger spin produces a larger drag, which makes the ball slow down more rapidly and thus decreases the distance it travels, but a larger spin produces a larger lift, which keeps the ball in the air for a longer time and thus allows it to fly farther. Experience tells us that the latter effect is predominant.

Attempts at Calculations

Various attempts have been made to calculate the trajectories for a ball hit at various velocities and spins, and those obtained look very much like those seen on the golf course. Such calculations are of little practical use to the individual golfer. The golfer had best approach this aspect of the game experimentally. He should bear in mind the effect of spin on lift and drag and should realize that the rate of spin imparted to the ball depends on the effective loft of the club face as it approaches the ball and on the velocity of the clubhead. The effective loft of the clubface will depend partly on the loft of the club used but also on how far ahead of the ball the hands are at impact. For a ball hit hard with a very lofted club, enough spin may be given to the ball to make it roll backward when it lands on the green.

When the Spin Axis Is Tilted

So far, we have considered the ball to have spin about a horizontal axis. It is the exceptional shot for which this axis is precisely horizontal. If the axis tips down to the right, the lift on the ball will be tipped to the right, and the ball will drift toward the right. This motion is called a slice for a right-handed golfer. If the axis tips down to the left, the drift will be toward the left. This motion is called a hook for a right-handed golfer. In order to hit the ball so that its spin has a horizontal axis, the clubhead should be exactly square to the intended line of flight and the clubhead should be moving exactly along the intended line of flight at impact. If the clubhead moves from the outside in or from the inside out with the clubhead square to the intended line of flight, the axis of spin will be tipped to produce a slice or a hook. If the clubhead moves along the intended line of flight but the clubhead is toed out or in so that the clubhead is not square to the intended line of flight, the axis of spin will be tipped to produce a sliced or a hooked ball. It is possible to hit a straight ball when the clubhead is toed in and moves from the outside in if the path of the clubhead and the orientation of the clubhead are chosen correctly. This would be a most difficult way to hit a straight ball and certainly would be recommended by no one.

Effect of the Wind

When there is wind, three velocities must be considered. One velocity is that of the ball relative to the ground and, equivalently, relative to the air through which the ball is moving when there is no wind. Another is the velocity of the wind relative to the ground. A third velocity is that of the air relative to the moving ball when there is a wind blowing. Two of these velocities determine the third, as shown in Fig. 8.2. In this figure the line OP represents the velocity of the ball relative to the ground. The radius of the circle represents the speed of the wind. If the ball is hit directly into the wind, the velocity of the air past the ball will be greater than if there were no wind. The velocity of the air past the ball will then be represented by the line AO. This velocity is the sum of PO and AP. If the ball is hit with the wind, the velocity of the air past the ball will then be the difference of PO and BP, which is BO. If the wind is directly across the intended path of the ball and from the left, the relative velocity of the air past the ball will be represented by the line CO. The line DO represents the relative velocity of the air past the

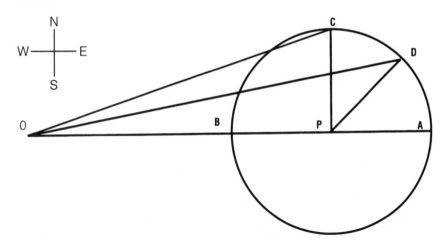

FIGURE 8.2. Diagram to aid in visualizing the velocity of the air past a ball moving through air when there is a wind. Let the line OP represent the velocity of the ball toward the east and let the radius of the circle represent the speed of the wind. If the ball is moving against an east wind, then the velocity of the air past the ball will be represented by a vector toward the west of magnitude and direction of AO. If there is a north wind, then the velocity of the air past the ball will be in the southwest direction given by the magnitude and direction of CO. If there is a wind from the northeast, then the velocity of the air past the ball will be toward the southwest given by the magnitude and direction of DO.

ball when the wind is quartered toward the ball from the left. The aerodynamic forces on the ball, the lift and the drag, are determined by the velocity of the air past the ball and not directly by the velocity of the ball or the velocity of the wind relative to the ground.

If we remember how lift and drag depend on velocity of the ball through the air, we realize at once that a ball hit against the wind will have greater lift and more drag, and one hit with the wind will have a smaller lift and less drag. The experienced golfer will keep his drives against the wind low and will hit them with less spin. He will hit his drives with a following wind somewhat higher than otherwise and will impart more spin to provide the necessary lift when the ball is moving more slowly relative to the air. For a 30-mile-an-hour wind, the drag, estimated on reasonable assumptions, may differ by a factor of 5 or more for balls hit against and with the wind.

We have seen that hooks and slices are produced by the same aerodynamic forces that produce the lift on a spinning ball. For a ball hit into the wind, besides the increase in lift there will be an increase in the force, producing a hook or a slice. Hooks and slices therefore will be accentuated for balls hit against the wind and will be diminished for those hit with a following wind. The wind condition must be considered as an important factor when a ball is to be hit with an intentional hook or slice.

Effect of a Crosswind

Let us next consider the flight of a ball hit into a crosswind. To make things easier to visualize, consider a ball somewhere on its way and moving east. The line OP in Fig. 8.2 represents the velocity of the ball relative to the ground. Assume a crosswind directly from the north. Its velocity will be represented by the line CP in the figure. The air will have a velocity past the moving ball represented by the line CO. If you imagine that you are riding on the ball, the wind appears to come from the northeast. The drag depends on the speed of the air past the ball, and this force will be in the direction of the air past the ball and not wholly in the direction of the flight of the ball. The ball will be slowed by the component of the drag along the path of the ball, as shown in Fig. 8.3. The component of the drag at right angles to the path of the ball is the force that makes the ball drift to the right in flight. If the speed of the wind is large, this force to the side may be fairly large compared to the drag and may make the ball drift an appreciable distance. In any case, this distance is much greater than would be expected if one were to estimate the drift produced by drag of the wind alone acting on the ball.

A Golfer Must Use His Judgment

Billiards is not played using a protractor to measure angles and a slide rule to make calculations. Nevertheless, the sometimes strange behavior of a billiard ball can be understood in terms of physical prin-

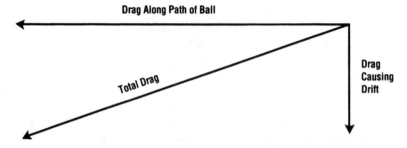

FIGURE 8.3. Diagram of the drag vector, the retarding force on the ball, when the ball is hit into a crosswind. The figure shows the components of the drag along and perpendicular to the path of the ball. This latter component of the drag causes the ball to drift at a right angle to its path.

ciples. Similarly, a golfer is not likely to carry an anemometer, a barometer, surveying instruments, and computing equipment to the golf course. In billiards the player looks over his shot, and thereby feeds into his brain the necessary information to make a reasonable decision on how to stroke the ball. I do not understand and I doubt that anyone understands just what the brain does in this process. We cover up our ignorance by simply stating that the player uses his judgment.

Similarly, in golf the player looks over his shot. He considers the slope and condition of the ground where he intends to have the shot land. He picks this area with the intention of making the next shot easier. He considers the strength and direction of the wind. He may pick a little grass and drop it to watch it drift in the wind. He evaluates possible hazards. He may even consider, in view of the current scores, the psychological atmosphere of the match. With all the information available, he uses his judgment and hits the shot. His judgment rests on his experience with similar shots in the past. If the shot does not come off as expected, the experienced golfer will maintain that if he were to have a second try he would attempt exactly the same shot. The inexperienced golfer would bemoan the fact that he had not thought the shot through as he should have and that on a second try, which of course he never has, he would make the correct shot.

I think that a thorough understanding of the physical principles underlying the various aspects of the flight of a golf ball will help the golfer to make correct judgments and thereby improve his game.

CHAPTER 9

When the Clubhead Meets the Ball

The Quality of the Shot Depends on the Golfer

One Saturday afternoon a friend told me as we were walking off the fifth tee that he was developing a habit that produced a certain aberrant flight of the ball and asked me to pinpoint his trouble. At that time, about all I could do was listen and express my sympathy. A couple of weeks later at about the same place on the course he again asked me to help him. His second request led me to do some fundamental thinking about what actually happens when the clubhead meets the ball. The action of the ball after the collision and the process during the collision must be determined by what the golfer does before the collision. The responsibility for the quality of the shot lies exactly where it belongs, with the golfer.

Description of the Collision

During the collision between the clubhead and the ball several things happen. The clubhead is slowed down, and the ball is sent off with a high speed at some angle above the horizontal with a high rate of spin. This happens in less than a thousandth of a second while the clubhead is moving less than an inch [3]. Such a short time makes it very difficult to observe what is happening during the collision. The force between the ball and the clubhead averaged over the time of the collision is greater than 3000 lb, and high speed photography has shown the ball to be considerably flattened against the clubhead [4]. The elastic properties of the ball come into play when the ball is compressed and then springs away from the clubhead. Frictional torque comes into play to set the ball spinning. We can only guess at the details of the collision, but we can use some physics to make an educated guess.

Momentum Is Conserved

We shall first apply the principle of the conservation of momentum to the collision between the clubhead and the ball. Since the clubhead is at the end of a somewhat flexible shaft, one may, to a fair degree of approximation, assume that the clubhead in its horizontal motion at the bottom of the swing acts as a free body. Consequently, the horizontal momentum of the clubhead before the collision must be the same as the sum of the horizontal momenta of the clubhead and the ball after the collision. The vertical component of the momentum is not conserved, because the golfer exerts a force on the clubhead by means of a pull on the shaft of the club. The principle of the conservation of momentum then accounts for the high speed of the ball after the collision and the slowing of the clubhead during the collision. A mathematical treatment of this and the following considerations may be found in the Technical Appendix, Section 3.

The Collision Is Inelastic

We shall now look at the effect of the elastic properties of the ball on the details of the collision. If the balls were perfectly elastic, meaning that no mechanical energy would be lost in the collision, things would be fairly simple. However, energy is lost in the change of shape of the ball and in the sliding friction of the ball on the clubface during the collision. This first loss can be treated in a fairly routine manner, but the second cannot. A collision in which mechanical energy is lost is called an inelastic collision.

Inelastic collisions have been studied since the time of Newton. If we drop a golf ball on a hard floor, it will rebound to a height somewhat less than that from which it was dropped. If no energy were lost in the collision with the floor, the ball would rebound to the same height from which it was dropped. Newton found an approximate rule describing this effect. He found the ratio of the speed with which the ball leaves the floor to that with which the ball approaches the floor to be practically a constant over a large range of speeds. The constant is called the coefficient of restitution. For a perfectly elastic ball, the coefficient is one. For a ball that does not rebound at all, such as one of putty, the coefficient is zero. Experience shows that this general rule may be extended to collisions where the ball rebounds at an angle from a smooth floor. Difficulties arise when the floor is rough and friction comes into consideration. The face of the golf club is grooved to ensure that the collision between the golf ball and the face of the club does involve friction, a condition necessary to produce spin on the ball, as we shall see. If we ignore the sliding friction, we may use Newton's rule to describe the elastic characteristic of the collision to a rough approximation. Such a description may be written in the form of a mathematical expression.

The Ball Slides and Rolls on the Clubface

We shall finally consider the effect of this sliding friction between the ball and the clubface. When the clubhead begins to make contact with the ball, the ball will begin to slide up the clubface, with the force between the ball and the clubface gradually increasing. The resulting frictional force on the ball will gradually give the ball a rolling motion, and when the ball is about to leave the clubface, the ball will be rolling without sliding if there has been enough friction. This action may be visualized by considering what happens when a bowling ball is released on the alley on its way toward the pins. The bowling ball first starts to slide. Since there is a frictional force between the ball and the surface of the alley, the ball does two things: It slows down, and at the same time it starts to roll. After the ball has gone some distance toward the pins it will be rolling without sliding. This motion of the bowling ball is much simpler than that of a golf ball, since the bowling ball barely flattens against the alley, while the golf ball flattens an appreciable amount against the clubface. However, the overall effect is similar. The golf ball has acquired a high rate of spin about a horizontal axis by the time it leaves the clubface. The dynamics of this process allow us to write two more mathematical expressions involving the rate at which the ball is set spinning. One of these expressions is uncertain to some extent because of the difficulty of bringing the flattening of the ball into consideration with precision. The details of these mathematical developments are found in the Technical Appendix, Section 3.

Any golfer who is skeptical about the fact that a golf ball is set spinning when it is hit by a club may set up an experiment to observe this effect. Hang a ball in a doorway on a string made of two threads using some adhesive tape of some kind. Two threads are needed because a single thread will continue to unwind. The ball must be hung so that it almost touches the floor. The ball is then tapped with a putter with the blade held at various angles. When the blade is held square to the line along which the putter is moved in tapping the ball, the ball will recoil without spin, but when the blade is at an angle to this line the ball will recoil and be set spinning about a vertical axis. For a given speed of the putter the rate of spin will be determined by the angle of the blade.

Clubface Orientation

As we proceed with this discussion we shall need a word to describe the orientation of the clubface as it collides with the ball. A line perpendicular to a surface is called the "normal" to the surface. If you use one eye to look at its image in a mirror, the line of sight is normal to the surface of the mirror. The orientation of the clubface may be specified, then, by giving the direction of the normal to the clubface.

The Loft of a Club

We shall next define the loft of a club. A club usually has two planes built into it. One plane is the face of the club, and the other is the base of the club. The dihedral angle between the two planes, which is less than a right angle, is then the complement of the loft of the club. As an example, consider a "standard" seven iron. The angle between the base of the club and the face of the club is sixty degrees. The complement of sixty degrees is ninety degrees minus sixty degrees, or thirty degrees. The loft of a "standard" seven iron is thus thirty degrees.

The Effective Loft of a Club

The effective loft of a club moving toward the target and striking the ball will usually not be the loft of the club. The effective loft of the club will be the angle between the normal to the face of the club and the velocity vector of the clubhead in a vertical plane containing the ball and the target.

The effective loft of the club, EL, depends on several factors. If the golfer is using a club with a flexible shaft, the effective loft of the club will depend on its flexibility and on any peculiarities in swinging the club. We shall not attempt to analyze this factor. Golfers have expressed their sentiments in this matter by saying that they want stiff shafts. One aspirant for the tour said, "The stiffer the better." The effective loft of the club depends on whether the clubhead at impact takes a divot. If it does, the velocity of the clubhead at impact will be at some angle below the horizontal, and the effective loft will be decreased by this angle. When these two factors, the stiffness factor and the divot factor, are omitted from further discussion, any effect of these two factors may simply be added to or subtracted from the effective loft EL.

Neglecting these factors, the effective loft of a club may be shown to be given by the expression $EL = L + \alpha(i) - \beta(i) - \gamma$, where L is the loft of the club, $\alpha(i)$ and $\beta(i)$ are α and β when the clubhead is in contact with the ball, and γ is the backswing angle of the arms (see Fig. 2.3). The angles $\alpha(i)$ and $\beta(i)$ will be determined by a calculation for each individual swing. We shall see that these angles depend on various characteristics of each swing.

The effective loft of the club is one factor in determining the spin of the ball, its rate of rotation. When a club with zero effective loft hits the ball, the ball leaves the clubface with no spin. When the effective loft is greater than zero, there is a complicated interaction between the club and the ball that depends on the friction between the two surfaces, that of the ball and that of the clubface. It also depends on the elastic properties of the ball, on the moment of inertia of the ball, and on the clubhead speed.

The Spin of the Ball

When the ball leaves the clubhead after a faultless stroke, it is spinning at a high rate about a horizontal axis with the top of the ball moving toward the golfer relative to the center of the ball. If the golfer has trouble with his swing, the axis of the spin may not be horizontal.

The Forces Called Drag and Lift

When the ball is on its way and there is no wind, the air exerts a force on it that may be resolved into two components. One component, called drag, may be described as a retarding force on the ball. This component is along the path of the ball and gradually slows the ball. The other component, called lift, is an upward force at a right angle to the path of the ball. Lift depends on the rate of spin of the ball and on its speed through the air. Both of these forces depend on the speed of the ball and on its rate of spin.

The length of a golf shot depends on many factors, drag being one of them. If the drag on the ball increases with its rate of spin, it must decrease as the ball is hit with less spin. Some of those writing in this field have suggested that while distance does depend on the speed of the ball, it may also be increased by hitting the ball with less spin, because this maneuver reduces the drag. But it is also known that the lift on a spinning ball keeps the ball airborne for a longer time and therefore allows the ball to go farther before it hits the ground.

Factors Affecting Drag, Lift, and Spin

It is of interest to consider factors by which the golfer may reduce the spin of a ball and thereby reduce the drag and lift of the ball. One way is to reduce the effective loft of the club. This may be done by choosing a different club with less loft.

A computer calculation was made for the standard swing with the wrist-cock angle at the top of the backswing increased slightly with the swing being otherwise unchanged. The effect of this change was to decrease the effective loft of the club. A similar calculation was made with the backswing angle of the arms slightly increased. The effect of this change was to increase the effective loft of the club. Another similar calculation was made with the torque on the system, TS, slightly increased. The effect of this change was to increase the effective loft of the club. Another calculation was made with the acceleration of the shift slightly increased. The effect of this change was to slightly increase the effective loft of the club. We know that increasing the effective loft of the club increases the rate of the spin of the ball and also increases the aerodynamic lift on the ball.

The following arrangement should help the reader to keep the results of these calculations in mind. The first column refers to the quantities changed in the calculations. The plus and minus signs indicate a slight increase or decrease in any of these quantities. The other columns represent the changes experienced by other quantities found in the calculations.

$\beta(0)-$	$\beta(i)-$	EL+	Spin+	Lift+
$\beta(0)+$	$\beta(i)+$	EL−	Spin−	Lift−
$\gamma-$	$\beta(i)+$	EL−	Spin−	Lift−
$\gamma+$	$\beta(i)-$	EL+	Spin+	Lift+
TS+	$\beta(i)-$	EL+	Spin+	Lift+
Al+	$\beta(i)-$	EL+	Spin+	Lift+

As an example, the third line above tells us that when the backswing angle of the arms is decreased, the wrist-cock angle at impact is increased, the effective loft is decreased, the spin is decreased, and the lift is decreased.

One computer calculation for the standard swing was particularly interesting. The wrist-cock angle at the top of the backswing was increased, and the backswing angle of the arms was decreased, both by five percent. The effective loft of the driver with a ten-degree loft turned out to be negative, and for the driver with a twelve-degree loft, it turned out to be almost exactly zero. Thus with these two changes, in the standard swing the ball would have no lift at all. This particular calculation helps in our understanding of how a swing that is serviceable for one golfer may be of no use to another. We have looked at four factors that affect the effective loft of a club. A professional golfer who was trying out a set of clubs perfectly matched to a seven iron, according to my method of matching as discussed in Chapter 11, reported that the clubs played well except for the one iron. He could not get the ball up with this particular club; the effective loft of the club seemed to be negative or essentially zero. I suggested that he swing the club as he would a seven iron. He found that with such a swing he had no trouble getting the ball up with the one iron. Calculations using the standard swing have shown that small variations from that swing produce surprising effects in what happens to the ball.

So far, we have made small changes in how the club is to be swung. Let us look at a swing where a different driver is used in the standard swing. Instead of having the professional swing his own driver, let us have him swing my driver. This means changing the three mechanical properties and the length of the club in the computer program. The effective loft with the change in the driver was increased about 18 percent when both drivers had a loft of ten degrees. It was about 11 percent when each driver had a loft of twelve degrees.

A golfer who is always trying to improve his swing and at the same time is always looking for a new and better set of clubs may have a problem involving so many variables that finding a solution is purely a matter of luck. Two remarks of Jack Nicklaus shed some light on how he chose his clubs. He advises, "I believe a golfer should use the same swing in all conventional shots." And after describing his driver in detail, he tells us, "I got to this through trial and error" [9].

The *D* Plane

Euclid, the famous geometer, says something to the effect that two intersecting lines determine a plane. The normal to the clubface and the line along which the clubhead is moving at impact intersect at the ball and therefore determine a plane. The line along which the ball leaves the clubhead also lies in this plane. We shall call this plane the *D* plane because it is descriptive of the collision between the clubhead and the ball.

Consider a collision for which the clubhead at impact is moving directly toward the chosen target and the normal to the clubface is directed to a point exactly above the target. The *D* plane for such a collision contains the target, and the plane is vertical. After the collision, the ball will be moving in this plane toward the target with the line of flight a little below the normal to the clubface. Let us have this type of shot in mind while we begin to discuss the mathematical expressions shown in the Technical Appendix, Section 3.

Equations Give Some Results

After the speed of the clubhead is given, the mathematical equations may be used to determine approximately what is happening after the collision is over. They will tell us the speed of the clubhead, the speed and direction of the flight of the ball, and the rate of spin of the ball.

The velocity vector of the flight of the ball after the collision is below the normal to the clubface at impact. For similar swings the angle between these two vectors increases almost directly with the loft of the club.

The rate of spin of the ball depends directly on the speed of the clubhead at impact. The golfer who can move the clubhead at a higher speed will put more spin on the ball, and consequently, the ball will experience a greater lift. The lift will be greater because it increases with the spin of the ball and it also increases with the speed of the ball through the air. With the greater lift, the ball will be held up longer, it will not fall as fast, and with the higher speed it will fly farther before it hits the ground. This is true even though the drag also increases with the speed of the ball.

The rate of spin of the ball depends directly on the speed of the ball as it leaves the club and almost directly on the angle between the normal to the clubface and the line of the flight of the ball as it leaves the club. For a given speed of the ball, the spin rate is about three times greater with a five iron and about five times greater with a nine iron than with a driver.

Collision Theory

The theory of collisions indicates that if a very light perfectly elastic ball is hit with a very heavy club with no loft, the ball will leave the clubhead with almost twice the speed of the clubhead before impact. This is an upper limit. For the ordinary ball and an ordinary club, the ball will be sent off at a slower speed. For a driver, five iron, and nine iron, the factors are 1.46, 1.30, and 1.12 for the three clubs, respectively, according to our theoretical results. The photograph of Bobby Jones swinging his driver [4] shows the ball's speed to be 1.4 times the clubhead's speed before impact. The agreement between 1.4 and 1.46 is better than we have any reason to expect because of possible experimental error in determining the speeds involved and also in determining the coefficient of restitution of the golf ball. The agreement suggests, however, that our theoretical results are not too far out of line.

We may use the theoretical results to calculate the rate of spin of the ball. Some experimental determinations of both the rate of spin and the speed of the ball are available [7]. These data are for the swing of a five iron by a professional golfer, and the exact orientation of the clubface at impact is not given. By using a 5 deg angle between the normal to the clubface and the line along which the ball leaves the clubface and a ball speed of 145 ft per second for a five iron, the theory gives a calculated rate of spin of 72 revolutions per second. For a five iron the measured spin rate for this ball speed averaged about 80 revolutions per second. Perhaps a difference of this magnitude should be expected as a result of the approximations made in setting up the original mathematical expressions.

The spin put on the ball when it is hit correctly, so that it rolls up the clubface during the collision, is such that it appears to rotate in a counterclockwise sense as seen by an observer facing a right-handed golfer. The ball is given what has been called underspin. A ball must rotate with underspin in order that the aerodynamic forces give the ball the necessary lift. If the ball is hit with the edge of the club above the center of the ball, the ball will be given topspin. Such a topped ball will have negative lift. This negative lift accounts for the quickly diving flight of a topped ball.

When the ball is hit out of deep rough and a layer of grass lubricates the contact between the clubface and the ball, there may not be enough friction to give the ball the usual amount of spin. Under this condition the ball may leave the clubface along a line closer to the normal than usual. Expert golfers have learned that a nine-iron shot from deep rough will not bite the green as does a shot with the usual amount of backspin.

The D Plane as a Practical Tool

The D plane for a golf swing contains the path along which the clubhead is moving at impact, the normal to the clubface, and the initial path of the ball after impact. The D plane also contains the aerodynamic lift force, since the lift force is perpendicular to the axis of spin and this axis is perpendicular to the D plane. A diagram of the D plane for a possible five-iron shot is given in Fig. 9.2. A copy of this diagram on a card will facilitate the reading of the following discussion.

It is suggested that a golfer take a five iron in hand as he reads this description of the use of the D plane. If the reader takes the usual stance with a five iron and swings the club directly in the direction of an assumed target with the clubface square to the target, neither toed in or toed out, then the D plane for the swing will be a vertical plane containing the velocity of the clubhead, the velocity vectors of the ball and the normal to the clubface, and the target. To illustrate this swing, the card representing the D plane should be held so that the line representing the clubhead motion points horizontally toward the target and the line representing the normal to the clubface points directly toward but above the target. For such a swing, in the absence of a crosswind, the ball will fly directly toward but above the target without a hook or a slice.

Next, let the reader open the face of the club so that the normal to the clubface points to the right and above the target. The club has been toed out. This is done while retaining the original intention of swinging the club directly at the target. To represent this swing of the club, the *D* plane is tilted to the right. The line representing the direction of the clubhead motion is still horizontal and points toward the target. The line representing the normal to the clubface points above and to the right of the target. The card shows that the path of the ball will also be to the right of the target but not as far to the right as the normal to the clubface. Since the lift force lies in the *D* plane, the lift force will have a component to the right, and this component will produce a slicing motion of the ball.

For a club that has been toed in, the *D* plane is tilted to the left, and the previous discussion applies except that "left" replaces "right" and the component of the lift force to the left will produce a hooking motion.

Many swings with a golf club are made in which the clubhead is not directed toward the intended target. For such swings, the line on the *D* plane card representing the direction of the clubhead motion, while still horizontal, must be directed either to the right or to the left of the target. If the swing is described by a vertical *D* plane, the ball will go to the right or to the left without hooking or slicing. The reason for this is that the horizontal component of the normal to the clubface also points along the direction of the clubhead motion. However, if for these cases where the clubhead motion is not toward the target, and the clubface is toed out or toed in, the *D* plane is tilted as before, and the ball will have a slicing or a hooking motion.

It is usually taught that a swing of a clubhead from outside in will produce a slice, but we see that a slice will result only if the swing can be described by a *D* plane tilted to the right, and if the *D* plane is tilted to the left, the ball will have a hooking motion in spite of the motion of the clubhead from the outside in.

The analysis involving the *D* plane may be used in diagnosing any swing producing an aberrant flight of the ball. Consider the problem facing a young man who played a round of golf with me. His tee shots consistently started off to the left and sliced back to the right. The slicing flight of the ball tells us that the *D* plane for his tee shots was tilted to the right; the aerodynamic lift force had a component to the right producing the slice. Since the ball started off along a line to the left of the target and this line is in the *D* plane, we know that the clubhead at impact was also moving along a line pointing to the left of

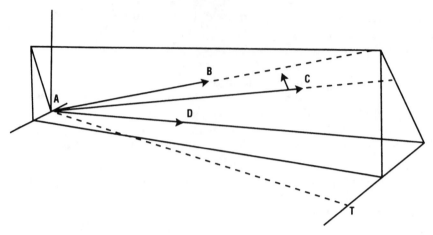

FIGURE 9.1. The plane containing three directed line segments represents the *D* plane for a slicing golf shot. In the diagram the ball is on the tee at A. The line AT represents the line from the ball to the target. The line AB represents the normal to the clubface at impact with the ball, the line AD represents the velocity vector of the clubhead after impact, and the line AC represents the velocity vector of the ball just after impact. For this shot the clubhead is moving from the outside in across the line from the ball to the target, and the ball will start off to the left of the target line and slice back toward the target line, since the lift vector is tilted to the right.

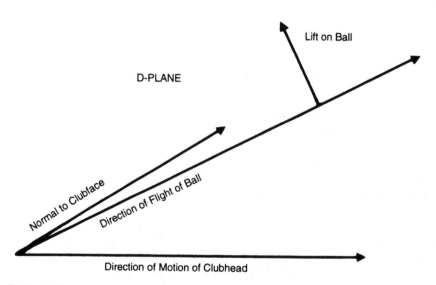

FIGURE 9.2.

the target. Since the normal to the clubface is in the tilted *D* plane and lies above the line along which the clubhead is moving at impact, the normal to the clubface must point to the right of this line. We have thus determined that the young man was swinging the clubhead from the outside in across the line to the target and the face of the club was open to the path of the clubhead. Our diagnosis was confirmed when on one of the tees, his driver cut through the grass below his ball. The grass was removed along a path pointing several degrees to the left of the green. A diagram of the tilted *D* plane for this kind of shot is shown in Fig. 9.1.

The analysis involving the *D* plane may also be used to design swings to produce intentional hooks and slices. The club must be toed in or toed out to tilt the *D* plane to the left or right. If the ball is to be kept in the fairway, the line along which the clubhead is swung must be pointed to the right of the target for the intentional hook and to the left for the intentional slice. Figure 9.1 shows how the *D* plane must be oriented for the intentional slice.

Many golf instruction books treat the subject of sidehill lies. When the ball lies lower than your feet, you are to play for a slice, and when the ball lies higher than your feet, you are to play for a hook. The *D* plane may be used to understand why. Consider a sidehill lie with the ball higher than your feet. If you stand to the ball with your usual stance, the normal to the clubface will naturally point to the left of the target. This will tilt the *D* plane to the left, and the ball will have a hooking flight. The golfer has two possible ways to correct this flight of the ball. He may, as the books advise, play for a hook and aim to the right of the target. The question is, How much to the right of the target should he aim? He may toe the club out a bit so that the normal to the clubface and the direction of the motion of the clubhead at impact indicate a *D* plane that is vertical and contains the target. There will be no hook for a vertical *D* plane, but the toeing out of the club will give the club more loft. Thus for a given swing the ball will not go as far as one might expect for a given club. Then the question is, How far will the ball go? In either case the shot requires considerable judgment.

In considering what happens when the clubhead meets the ball, we have assumed that the ball makes contact with the center of the clubface, at the "sweet spot." For such a shot there is no torque about the shaft of the club to produce a change in the orientation of the clubface. If the ball makes contact with the clubface off-center, the clubhead will turn under the torque produced by the large force between the clubhead and the ball. If the ball is hit too close to the toe of the club, the club will turn to a toed-out position before the ball leaves the clubhead, putting the normal to the clubhead farther to the right than was intended. The ball will then fade to the right.

Perimeter Weighting of Clubheads

The head of the club may be designed to decrease the turning of the club for off-center shots. The rotation of the clubhead for such shots depends inversely on the moment of inertia of the clubhead about an axis parallel to the shaft through the clubhead's center of mass. Perimeter weighting of clubheads increases this moment of inertia. It is instructive to compare the moments of inertia of a uniform rectangular clubhead to one of extreme dumbbell shape. The moment of inertia of the dumbbell-shaped clubhead will be at most three times larger than that for the rectangular clubhead for the same mass and the same length. The ball would be sent off from these clubheads for off-center shots at angles that may differ by a factor of 3, the dumbbell-shaped clubhead giving the smaller angles. For any practical perimeter-weighted club, the factor would be less than 3.

For the usual clubhead design, the moment of inertia may not be too much different from the rectangular clubhead. The usual perimeter-weighted clubhead will have a moment of inertia considerably smaller than the dumbbell clubhead. This rough analysis leads one to expect little noticeable effect in the use of perimeter-weighted clubs over ordinary clubs. Others [3] have looked at the effectiveness of perimeter weighting in more detail and have concluded that any practical perimeter-weighted club would have a moment of inertia less than 50% greater than the classical clubhead. This means that if a certain off-center shot with a regular club would deflect the ball 3 deg, the same off-center shot with a perimeter-weighted club would deflect the ball somewhat more than 2 deg. If you are a golfer who hits the ball near the sweet spot of the clubhead, you would probably not be able to notice any improvement in your shots by using perimeter-weighted clubs. The practical value of perimeter-weighted clubs does not seem to be well established.

Off-Center Shots

The problem of off-center shots is real, and many golfers who lack precision on their approach shots should look at the possibility that they are not hitting the ball at the center of the clubface. Stroboscopic photographs show that even accomplished golfers sometimes hit shots that cause the clubface to turn while the ball and clubface are in contact [4]. A golfer can examine his swing for this error in a very simple manner. He should obtain a small tube of watercolor paint, preferably a bright color, and apply a very small dab of paint to the ball as it is placed on the tee. The ball is oriented so that the clubface first makes contact with the dab of paint. After the shot, the clubface is examined to see where it made contact with the ball.

CHAPTER 10

The Harry Vardon Swing

What Did Harry Vardon Do?

Today's competent right-handed golfer knows that he must swing the club with a straight left arm. Our calculations show that any attempt to make the clubhead move with greater speed at impact with the ball by using helping wrist action and particularly by bending the left elbow actually decreases the clubhead speed at impact. My interest in the golf swing would have been limited to the straight-left-arm swing had not a friend informed me that in his reading about the history of golf he had learned that a British golfer of a former day, Harry Vardon, had "a habit he never foresook but compensated for—bending his left arm at the top of his backswing" [25].

Harry Vardon, 1870–1937, was one of the world's most famous golfers. He won the British Open Golf Tournament six times in the years from 1896 to 1914 and the U.S. Open in 1900. He played mostly in the hickory shaft gutta-percha ball era. He was known for hitting a long ball and keeping his shots in the air, where they would not be involved with chance difficulties on the ground.

Before the days of "skin games" and hustlers he played matches involving large sums of money. These matches were like prize fights. His advice to golfers was, "Any good shots and the bad are your own making." He was said to be a poor putter. He made the overlapping grip popular; though it is known as the Vardon grip, it did not originate with him.

Others tried to use the Vardon swing, but with little success. Vardon could not explain what he did, and it is unlikely that even he knew exactly what he did in his swing.

My friend told me he had tried the Vardon swing with the bending of his left arm and thought that he could hit the ball farther using this swing. My response was that this type of swing was one step on the way toward being similar to the action of a bull whip, which could be thought of as a large number of short, flexibly connected rods. My friend knew of my interest in calculating golf swings on the computer. He asked me to study the Harry Vardon swing to see whether possibly it could be true that his swing would produce longer shots than the straight-left-arm swing when the swings were otherwise the same.

The Three-Rod Model

The model representing the Vardon swing would be a three-rod model; the golfer's arms would be considered as two rods with the club being the third. The technical aspects of the three-rod model are too involved to be presented here.

After setting up the three differential equations and writing a suitable computer program for solving them, preliminary calculations showed that the Vardon swing did indeed give greater clubhead speed at impact with the ball.

At this stage in my study, I had no idea how the lower part of the arm in my model should be managed during the downswing. If in my calculations I allowed a torque to act on the lower arm, a torque similar to the one that acts on the club in the latter part of the downswing, the clubhead would swing in above the ball. However, I knew that a bent left arm straightened early in the swing produced a lower clubhead velocity.

Some Practical Experience

I felt I needed some practical experience using the Vardon swing. I used a seven iron, which with a straight-arm swing I usually hit about 135 yards. I tried various versions of a possible Vardon swing with very mixed results; but one swing sent the ball sailing as I had never been able to do with a seven iron. I stepped off the distance to the ball and found it at 190 paces, which calibrates for me to be somewhat over 170 yards. The ball was hit into rough, and there was no wind affecting the flight of the ball. I had no idea what I had done in that swing.

We Can Guess What Vardon Did

We can only guess what Harry Vardon did with his left arm during the downswing. We know from geometry that he needed an almost straight left arm when the ball was hit; an arm seriously bent at impact would at least top the ball and perhaps miss the ball altogether. While we know that no action of the wrists is needed to uncock the wrists in the downswing, calculations show that the centrifugal torque in the rotating system on the lower arm is not adequate to straighten the arm by the time of impact.

Calculations Show That Vardon Had Something

Of the many possible ways Vardon could have straightened his left arm in the downswing, I chose to try the one in which the angle of bending, the angle between the extension of the upper arm and the lower arm, decreased to zero with a constant negative angular acceleration. Calculations were made for various angles of bending at the elbow in our model. This angle of bending of the lower arm has a zero angular velocity at the beginning of the downswing and gradually increases its negative angular velocity during the downswing. A golf ball dropped from rest has an essentially constant acceleration. These calculations showed that the clubhead velocity at impact became greater as the elbow was bent through larger and larger angles at the top of the back-swing. If at the top of the backswing the elbow was bent through the possible angle of 100 deg, the clubhead velocity at impact was about 50% greater than that for the same swing with no such bending.

We must not assume that the swing just described is the ultimate Harry Vardon swing. A calculation in which the constant angular acceleration of the angle of bending of the left arm was delayed 1/20 of a second into the downswing gave a clubhead speed 60% greater than the standard swing.

His Swing Is A Success

Without knowing exactly what Vardon had done, and without much practice, I began to use my approximation to the Vardon swing and found that I was able to increase the length of my drives, losing little if any directional precision. This increased length was noticed by one of my weekend golfing colleagues. He asked me what I was doing, whether I was into "monkey glands or something."

A golfer who is satisfied with the length of his drives using the straight-arm swing may look at the Vardon swing as an interesting bit of golf history. Those golfers who are always trying to be a little longer off the tee and are bothered by the consequent inaccuracies may discover, with practice, something close to what Vardon did in his downswing. Of all the possible ways of straightening the left arm in this type of swing, Harry Vardon must have found one that served him well and allowed him to become almost invincible in the years of his triumphs.

The Vardon Swing Not Recommended for Short Irons

I should offer a caveat that the computer's results are based on a simplified model of the swing, and any results of the calculations should be accepted only as approximations. I see no reason for using the Vardon swing on any of the short irons, since they are used with a straight left arm to achieve the ultimate in precision.

CHAPTER 11

The Matching of Clubs

In Search of Better Equipment

Years ago I read an article, probably in *Golf Digest*, on the matching of clubs that contained a statement that it appeared to be beyond the powers of modern technology to devise a scheme for the perfect matching of golf clubs. Later, I found myself stranded in an airport with nothing to do, and this statement came to mind. I started work on the problem of matching clubs, and before I boarded my plane, I had a glimmer of how to proceed with the general problem of perfect matching. The basic idea was to find a way to distribute mass (slugs) along the inside of the shaft. It was a long hard row to hoe before I was able to design a satisfactory set of perfectly matched clubs.

I thought perfectly matched clubs should all swing the same, and therefore a golfer would need to learn only one swing. By perfecting this swing he should be able to increase the precision of his game. I found, however, that most golfers were interested in clubs that would give them greater distance on the course and did not think in terms of possible increased precision.

There is no way of predicting how a person will react to a new situation. A young man with whom I happened to be playing noticed my club design. I explained what I had done and offered to let him try my five iron. He accepted my offer and hit an extra ball on a par five hole. When we found his ball (which was in the middle of the fairway), I asked him what he thought of the club. After estimating the distance from the tee, he replied: "Who in their right mind would want a five-iron that hits the ball 250 yards?" His assessment of my five-iron was most unusual.

In this chapter and in the Technical Appendix, I present enough detail of club set design so that anyone interested in finding whether the use of such a set of clubs would help him to improve the precision of his game can make or have made a set of clubs exactly to his liking, perfectly matched or otherwise.

The Need to Understand Dynamic Parameters

Any serious discussion of the matching of clubs requires an under-standing of the mechanical properties of a club described by its three dynamic parameters. These dynamic parameters determine how the club swings in the hands of the golfer. They are quantities that may be determined very precisely for a given club. Such a discussion of neces-sity becomes somewhat technical, but for an understanding of the matching of clubs this cannot be avoided.

One of these dynamic parameters is the total mass M of the club. Another is the first moment, S, about the wrist-cock axis, which is taken here to be 5 in from the grip end of the club. The third dynamic parameter is the second moment, or the moment of inertia, I about the same axis. A discussion of the definitions of these parameters and the methods of finding their values for a particular club can be found in the Technical Appendix, Section 1.

These three dynamic parameters, M, S, and I, are each related to certain aspects of the feel of a club in the hands of a golfer. The feel related to each parameter may be experienced separately. The feel related to the mass M may be experienced by simply hefting the club. The mass of the club is proportional to its weight. The feel related to the first moment may be experienced by holding the club in a hori-zontal position using the normal grip on the club. The torque tending to pull the head of the club downward is proportional to S, the first moment of the club about the wrist-cock axis. The feel related to the second moment, I, about this same axis may be experienced by waggling the club in the usual manner when the club is held in the vertical position. The torque experienced by the golfer in waggling the club at a certain rate is proportional to I. A golfer may find it of interest to feel the different S and I of the club when the axis is chosen near the head of the club instead of at the wrist-cock axis. Since these two parameters depend on the distribution of mass in the club about an axis, the difference in feel involving S and I about these two axes, one near the midpoint of the grip and the other near the head of the club, will be quite apparent.

Feel of Club Depends on Parameters

The differential equations of motion that result when Newtonian mechanics is applied to the study of the swing of a golf club show that these three dynamic parameters of the club are involved in the motion of a club under the forces and torques applied by the golfer. Newton's third law tells us that the forces and torques that the golfer feels during the swing are equal and opposite to the forces and torques that he applies to the club. Thus for all clubs of a set to feel the same in identical swings, they must have all three dynamic parameters respectively the same.

There have been various attempts to produce sets of clubs in which all clubs feel the same to the golfer. A trivial solution to the problem is to make a set of clubs all the same length with the same mass and the same mass distribution. Clubs of such a set would be perfectly matched, since they would have identical dynamic parameters and would feel the same. However, such a set of clubs would be rejected by most golfers because golfers are used to clubs of various lengths. Practical clubs vary only slightly from standard lengths. A standard five iron is 37 in long, and other clubs vary from this length in half-inch steps. Thus a nine iron would be 35 in long.

Swing-Weight Matching Is Only Partial Matching

A partial solution to the problem is obtained through swing-weight matching. In such matching, the three dynamic parameters vary smoothly from club to club. There is a very slight difference in feel between successive clubs in a swing-weight matched set, but the extreme clubs of the set do feel perceptively different to the golfer. A two iron from such a set does not feel the same as a nine iron.

Clubs of the same swing-weight have the same first moment about a point at a particular distance from the grip end of the club. Most of the balances found in pro shops determine the first moment of the club about a point 12 in from the grip end of the club. The 12 in distance is completely arbitrary.

The Lorythmic Swing-Weight Scale

While the swing-weight of a club may be expressed in any system of units, a curious system called the Lorythmic scale has become standard [7]. Swing-weights labeled C extend from 220 to 240 ounce inch, those labeled D extend from 240 to 260 ounce inch, and those labeled E extend from 260 to 280 ounce inch. Swing-weights under each label are divided into ten parts. Thus a club with a swing-weight of 246 ounce inch would carry the designation D-3.

There have been questions over the years whether matching by swing-weight is the best possible way to match clubs [7]. There seems to be no basis, other than custom, for matching clubs this way. Perhaps for a given golfer there are other ways of matching clubs that would be better for him.

A Possible Solution to the Matching Problem

It is not difficult to design a set of clubs in which one dynamic parameter is the same for all the clubs. There is no physical basis for thinking that clubs matched according to total mass or first moment or second moment would feel the same to the golfer swinging them. As I have shown, physical theory indicates that to have clubs feel the same they must be matched in all three dynamic parameters.

One attempt to match clubs in the three dynamic parameters has been partially successful. Clubs designed by this method have a single slug placed at some position within the shaft of the club with its mass and its position different for each club. The mass of the clubhead is also different for each club. The mass of the slug and the mass of the clubhead are chosen so that all clubs of the set have the same dynamic parameters. The application of this method leads to difficulties in the design of acceptable sets of clubs. However, perfectly matched sets of clubs were on the market at one time.

When one examines a set of these clubs, the most noticeable feature is that they are not of standard lengths. When the clubs of such a set were compared with the corresponding clubs of standard lengths, it was found that the two iron was 1.25 in longer and the nine iron was 2.1 in longer than standard. When one remembers that a set of irons can be perfectly matched by making all the clubs the same length, it appears that this deviation from standard lengths is about one-fourth of the way to the condition of equal lengths. A careful theoretical study of the design of these clubs shows that indeed this variation from standard lengths is needed to partially overcome difficulties inherent in this method. As a practical matter, some variation from perfect matching may be allowed in the matching of clubs, as long as this variation is not detectable by the golfer. However, in the advertising of the clubs under discussion, perfect matching as indicated by instruments was emphasized as a special virtue.

A Method for Perfect Matching

Another method of matching clubs in the three dynamic parameters is obtained in answer to the question, "What minimum distribution of mass along the shaft of a club will suffice, with the proper choice of clubhead mass, for exact matching of all clubs of a set?" The problem posed in this question has been shown to have a unique solution. The set can have exact matching if two slugs of different masses are placed at particular positions within the shaft of each club. One slug is placed at the point halfway between the wrist-cock axis and the center of mass of the clubhead, and the other is placed at the wrist-cock axis. The details of the solution to this problem are shown in the Technical Appendix, Section 2.

This second method of exact matching in the three dynamic parameters can be applied to clubs of standard lengths or any variation from these lengths desired by the golfer. In this method one of the shorter clubs is chosen as the master club of the set. All clubs longer than the master club can be matched to it exactly, while clubs shorter than the master club can be matched to it only approximately. In practice, the golfer chooses a master club, probably a seven iron, that feels just right for him. This club is chosen with care, because all the longer clubs of the set will swing exactly like the master club if the matching is perfect and the swing is the same.

The approach to the design of the other clubs of the set is indirect. It is best to start with club parts (clubheads, connectors, shafts of the correct lengths, and grips) for these clubs. The clubheads should be weighed, and the clubs should be assembled temporarily. The clubs, including the master club, should then have their three dynamic parameters determined according to the method described in the Technical Appendix, Section 1. One may also start with a swing-weight matched set of clubs and disassemble them when the clubhead masses are to be changed.

Let us assume that the two iron is the longest club of the set. Equations (9) and (10) of the Technical Appendix, Section 2, are used to calculate the mass $A(2)$ that is to be added to the clubhead and the mass $m(2)$ of the slug to be put at the midpoint of the shaft of the two iron. The quantity $A(2)$ will most likely be negative, and mass will need to be removed from the clubhead. After the slug of mass $m(2)$ is placed within the shaft and mass $A(2)$ is added to the clubhead, the total mass of this club will be increased somewhat. The resulting mass of this longest club will then be the mass of each club of the set. A slug must be placed within the shaft of the master club at the wrist-cock axis to bring its mass up to the mass of the longest club of the set. Notice that the longest club of the set does not necessarily have a slug at the wrist-cock axis.

The other longer clubs of the set, in this case the three, four, five, and six irons, will have their masses A and m calculated as for the two iron. For each of these intermediate clubs a slug must be placed at the wrist-cock axis to bring the total mass of the club up to the mass of the longest club of the set. This procedure, when carefully applied, will ensure exact matching of these six longer clubs of the set in all three dynamic parameters.

When the design is completed, the clubs may be disassembled and the clubheads adjusted to their correct masses as determined by the calculations. Slugs of the appropriate mass m for each club are placed within the shafts at the midpoint between the wrist-cock axis and the center of mass of the clubhead. Slugs of the correct mass are also placed within the shaft of each club at the wrist-cock axis to bring the total mass of each club to the desired value. The slugs may be secured within the shafts using epoxy cement. The clubs are then reassembled in their completed form. The three dynamic parameters of the completed clubs may be determined experimentally to find how closely they check with the design values.

The moments of clubs may be visualized by plotting them as in Fig. 11.1. The first moment about the wrist-cock axis is plotted as the ordinate, and the second moment is plotted as the abscissa. The moments are expressed in the centimeter-gram-second system of units.

In this figure the points within the triangles represent the experimentally determined moments of a high-quality set of swing-weight matched clubs. This set was a Tommy Armour set produced by P. G. A. Victor, and it was chosen because it included a one iron. The points for most of these clubs fall near a straight line that slopes up and to the left; this is characteristic of swing-weight matched clubs. It appears that the designers of this set of clubs considered the nine iron and the pitching wedge to be specialty clubs, since these clubs are not matched with the others of the set. The swing-weights of the one, four, and seven irons are D-2, D-3, and D-4, respectively. The number to the right of each point indicates the mass of the club in grams.

The points for a set of clubs matched according to the first moment would fall along a horizontal line in the figure, and those for a set matched according to the second moment would fall along a vertical line. A set of clubs perfectly matched in all three dynamic parameters would be described by a single point in the figure.

An Example of Perfect Matching

As an example of perfect matching that may be achieved by the method shown in the Technical Appendix, a calculation was made to match the longer clubs of the set illustrated in Fig. 11.1, the Tommy Armour clubs, to the eight iron of the set as the master club. So that the resulting values of S and I can be shown in the figure without the superposition of points, the eight iron was modified by decreasing its clubhead mass by 1 g. The result of the calculation allows perfect matching of all the clubs one through eight with the following values of the dynamic parameters: $M = 477.7\,g$; $S = 25,960\,g \cdot cm$; $I = 1,952,700\,g \cdot cm^2$. All of the clubs of the matched set are represented by the single point outlined by the square in Fig. 11.1.

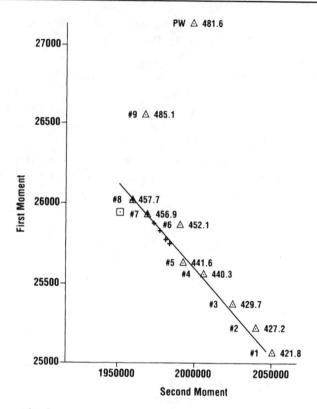

FIGURE 11.1. The first and second moments of a golf club plotted as ordinate and abscissa, respectively. The points within the triangles represent the moments of a set of swing-weight matched clubs carrying the name "Tommy Armour," produced by P. G. A. Victor. Notice that these points fall along a straight line sloping up to the left on the diagram except for the nine iron and the pitching wedge. The number to the right of each of these points indicates the mass of the club in grams. The calculation for a perfect matching of the Tommy Armour set to the eight iron, made 1 g lighter, is shown by a single dot in the square. The example of the same Tommy Armour set with relaxed matching is shown by the crosses (+) when the seven iron was taken as the master club.

The modifications that must be made in these clubs in the matching process are shown in Table 11.1. Here A is the mass to be added to the clubhead, m is the mass to be placed midway between the wrist-cock axis and the center of mass of the clubhead, and μ is the mass to be added to the club at the wrist-cock axis.

This first example is presented more as a design exercise to show that three dynamic matching can indeed be achieved rather than to show the design of a practical set of clubs. These clubs would all swing and feel like an especially heavy eight iron.

Table 11.1.

Club	A	m	μ
#1	−35.9 g	91.6 g	0.0 g
#2	−32.1	81.0	1.6
#3	−27.0	67.4	6.5
#4	−19.9	47.0	8.3
#5	−15.7	39.0	12.8
#6	−11.0	23.8	12.8
#7	−5.6	11.4	15.0
#8	−1.0	00.0	21.0

A Example of Relaxed Matching

Except when perfect matching is used as a basis for an advertising program, there is no need for clubs to be perfectly matched. A set of clubs will serve as well as one that is perfectly matched when the golfer using them cannot detect by feel or appearance any relaxation from perfect matching. However, relaxed matching offers some advantage in club design: both A and m for the longest club of the set can be made smaller, resulting in a more massive clubhead for better momentum transfer to the ball and a smaller total mass for all the clubs of the set.

The fractional difference that can be detected in the masses, the first moments, and the second moments between two clubs depends on the sensitivity of the person making the comparison. I find it fairly easy to detect the difference between a nine iron and a two iron in a swing-weight matched set for each of these parameters. I doubt that I can detect with certainty that there is such a difference between a three iron and a four iron. Proceeding in this way, I find that for me fractional differences of 0.015 for the mass of clubs and 0.007 for the moments of clubs is on the edge of being detectable. The relaxation from perfect matching probably should not be greater than these fractional differences.

The equations used in the exact matching of clubs, as developed in the Technical Appendix, may be modified to relax the matching of any two clubs to any preassigned degree. A second example is shown in Fig. 11.1, using crosses, for the same Tommy Armour clubs, where the seven iron is taken as the master club and the matching for S and I is relaxed to 0.7% or less and that for M is relaxed to 1.3% or less over the clubs of the set from the one iron to the eight iron. The results of the calculation for this set of clubs are shown in Table 11.2. Here A is the mass to be added to the clubhead; m is the mass to be placed midway between the wrist-cock axis and the center of mass of the clubhead; M is the total mass of the club; S and I are the first and second moments,

TABLE 11.2.

Club	A	m	M	I	S	p	q	r
#1	−25.5	66.6	462.9	1 984 000	25 770	0.007	−0.007	0.013
#2	−22.3	57.5	462.4	1 982 000	25 790	0.006	−0.006	0.012
#3	−18.6	47.9	459.0	1 978 000	25 850	0.004	−0.004	0.005
#4	−13.1	33.8	461.0	1 974 000	25 900	0.002	−0.002	0.009
#5	−10.5	28.2	459.4	1 970 000	25 950	0.0	0.0	0.005
#6	−5.6	12.7	459.1	1 970 000	25 950	0.0	0.0	0.005
#7			456.9	1 970 000	25 950			
#8	0	0	457.7	1 959 000	26 040	−0.006	0.004	0.002

respectively, about the wrist-cock axis; and p, q, and r are, respectively, the fractional differences between I, S, and M of the various clubs and the master club. All units are in the cgs system.

Several comments may be made concerning this second example. The values of p, q, and r show that the variations of the clubs from the master club are likely to escape detection for all but the most sensitive individuals. These clubs will swing like the seven iron, and they will feel like the seven iron. The masses to be removed from the clubheads are consistently less than those in the first example, and the masses to be added to the midpoints of the shafts are also considerably less. There is no need for masses to be placed in the shafts at the wrist-cock axis, and the masses of the clubs are less than in the first example by 14.8 g for the one iron to 20.8 g for the seven iron. If a golfer is strong enough to swing a normal seven iron, he should have no trouble with these clubs.

A Design for a Particular Individual

This second example is presented to show what may be accomplished in the matching of clubs when the requirement of exact matching is relaxed somewhat. A third example to be presented is to show what can be done in the design of a set of clubs for a particular golfer. Lightweight (response) shafts of the standard lengths were used, and the clubhead mass of the master club, the seven iron, was chosen by experiment to satisfy the golfer who prefers to swing heavier clubs.

The head of a seven iron with a lightweight shaft was machined so that by adding various masses to the back of the clubhead the mass of the club could be varied from 400 to 500 g without changing the position of the center of mass of the clubhead. By actual tests it was found that a club mass of 450 g gave the best results for this golfer. The factors influencing the choice of clubhead mass have not been pinned down. Momentum transfer to the ball is not involved, since for the clubhead masses used, the change in the momentum transfer is but 0.04% per gram.

A professional golfer working with this seven iron and a four iron matched to it found that a club mass of 435 g was optimum for him. When these clubs were used with more massive heads, the angular accuracy of his shots was reduced in a very pronounced and surprising manner. From these tests one is led to suspect that for a golfer with a well-established swing, the clubhead mass is a critical factor in the overall quality of his shots.

This set of clubs consisted of the three iron through the nine iron. In the design of this set the matching with the master club, the seven iron, was relaxed for the three iron and the four iron. The eight iron and the nine iron were matched very closely to the master club in both the mass and first moment, but the second moment was allowed to deviate from perfect matching by a significant amount. The design parameters of this set of clubs are shown in Table 11.3. The column headings are the same as used previously except for Mh, which indicates the club-head mass. All units are in the cgs system.

The clubhead masses of this set of clubs were adjusted to the values shown in Table 11.3, and slugs of masses m were inserted in the shafts at the specified positions, all with suitable precision. These slugs could not be the point masses given by the theory, but a separate calculation showed that the slugs could be distributed over a length of 20 cm with the center of mass at the specified positions with insignificant error.

The values of S and I for these clubs are shown in Fig. 11.2. Since these clubs had lightweight shafts, a greater fraction of the mass of the club was placed in the clubhead. Thus values of S and I are seen to be greater than for the usual clubs. The professional golfer for whom these clubs were designed was very pleased with them at the 450 g mass, and he said that the 435 g four iron matched to the 435 g seven iron was the best four iron he had ever used. Since it is difficult to distinguish with certainty between subjective and objective evaluations of clubs, any

TABLE 11.3.

Club	M	S	I	Mh	m	μ	p	q	r
#3	450.4	27 101	2 135 000	253.6	56.1	0	0.005	−0.005	0.002
#4	450.3	27 101	2 135 000	266.8	40.5	4.0	0.005	−0.005	0.002
#5	450.0	27 237	2 124 000	276.9	35.3	0	0	0	0.003
#6	450.4	27 237	2 124 000	292.2	17.4	4.6	0	0	0.003
#7	451.2	27 237	2 124 000	307.0	0	8.7	0	0	0
#8	450.7	27 253	2 094 000	314.0	0	4.0	0.014	0.001	0.001
#9	451.4	27 210	2 057 000	320.3	0	0	0.032	−0.001	0

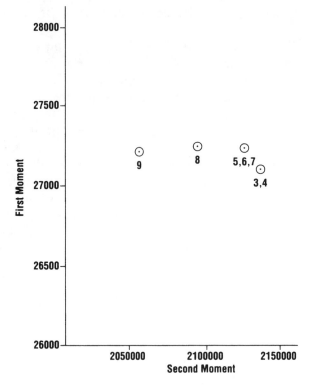

FIGURE 11.2. The plot of the first and second moments of the set of clubs described in Table 11.3. The seven iron was chosen as the master club. The longer irons were held to within 0.5% of perfect matching, while the shorter irons, eight and nine, deviated 1.4% and 3.2% from perfect matching.

further remarks on the quality of these clubs would be out of place here.

All clubs of a three-dynamic-parameter matched set will have the same mechanical properties, and they will swing like the master club of the set. They should all be swung with the same swing the golfer used when choosing the master club. Another professional golfer using a set of these clubs found he had difficulty getting the ball up when he used the long irons. When it was pointed out to him that habit induced him to swing the long irons as he had learned to swing the long irons of a swing-weight matched set and that he was not swinging them as he would swing the master club of the set, he found that indeed the clubs should all be swung as the master club, and he had no further difficulty hitting the ball properly with the long irons. No one using clubs matched in the manner described here has found that he was unable to obtain adequate distances with them.

More Research Needed

Many questions related to the use of clubs matched this way remain to be studied. This field of research is open to those who may be inclined to find whether a better game of golf might be played with clubs matched so that golfers cannot detect any difference except in loft and length of shaft.

CHAPTER 12

The Flexibility of Shafts

The Lateral Flexibility of Shafts

We have seen that a golf club has three dynamic parameters, the mass and the first and second moments about the wrist-cock axis, which may be measured with some precision. The golf club has another characteristic of interest to the golfer. This is the lateral flexibility of the shaft of the club. This characteristic is related to the properties of the material of which the shaft is constructed. Early clubs had wooden shafts. Next came tapered tubular steel shafts. Recently, clubs have been available with lightweight tubular shafts of high-strength steel and also with shafts made of aluminum, titanium, and graphite. Shafts may be made with various degrees of flexibility. At the present time the understanding of the effects of lateral flexibility of the shaft on the character of the golf stroke seems to be clouded by considerable perplexity.

Shaft Clamped in a Vice

When any golf club is clamped in a vice and the clubhead is pulled to one side and released, the club will oscillate with gradually decreasing amplitude. The frequency of the oscillation is influenced mainly by the stiffness of the shaft of the club, the mass of the clubhead, and the position of the point on the shaft where it enters the vice. A stiffer shaft produces a higher frequency, a greater clubhead mass produces a lower frequency, and a shorter length of shaft extending from the vice produces a higher frequency. A calibrated strobe light is useful in measuring the frequency of these oscillations. Both woods and irons when clamped in this way will have frequencies in the range of four to five vibrations per second. In a third of a second, about the time for a downswing, a club with a frequency of 4.5 vibrations per second, when clamped in this way, will make 1.5 complete vibrations.

A New Model Needed

The purpose of this investigation of the effect of the shaft flexibility on the golf stroke is to try to understand why clubs flex as they do during golf strokes and what effect the flexing of the club has on the speed of the clubhead as it hits the ball and also on the effective loft of the club. We shall make calculations using three simultaneous differential equations to try to find what factors must be assumed to be acting during the downswing so that our calculations will fit what we have found concerning the flexing of the shafts during the actual swings of clubs. This procedure will involve a modification of our model of the swing to include a third angle variable besides the angles of the arms and wrist cock used previously. This angle D will be the angular displacement of the clubhead, measured from the wrist-cock axis, that results from the bending of the shaft at any time during the downswing. The angle D will be measured in the same sense as the wrist-cock angle. The angle D will then be positive when the clubhead is left behind in the motion of the downswing. The differential equations are discussed in the Technical Appendix, Section 4.

Photographs Show Flexing of Shaft

The experimental angles D shown in Fig. 12.1 were determined by examining one of the 35 mm slides similar to that used in producing the enlargement in Fig. 2.1 of Chapter 2. This photograph was made at another time and of another professional golfer. The positions of the images of the three bits of tape on the shaft of the club for each flash of the Strobotac, when projected on a large sheet of paper, were easily seen not to lie in a straight line. The instantaneous angular deflection of the clubhead, measured from the position it would have had if the shaft had been inflexible, could be determined from measurements of how much the images of the three bits of tape deviated from a straight line. The values of the angle D for the various times into the downswing are plotted in Fig. 12.1. The crosses and dots indicate values obtained from two separate and independent determinations of the angles D. The scatter of points on the graph simply indicates the difficulty in making the measurements and gives some idea of the uncertainty of the measurements. However, the data have certain characteristics that need to be explained to obtain an understanding of the flexing of the shaft during a downswing.

How the Shaft Flexes

As seen in Fig. 12.1, the shaft bends, leaving the clubhead behind at the beginning of the backswing. This is a natural effect; the shaft bends as it does because it must exert a force on the clubhead to produce its acceleration. The shaft becomes straight at about 0.29 s into the downswing; the club is about horizontal at this time. The shaft then bends toward the target in ever increasing amounts until the ball is hit. There is a curious hump in the curve from about 0.16 to 0.23 s. This hump seems to be a real effect in this particular swing.

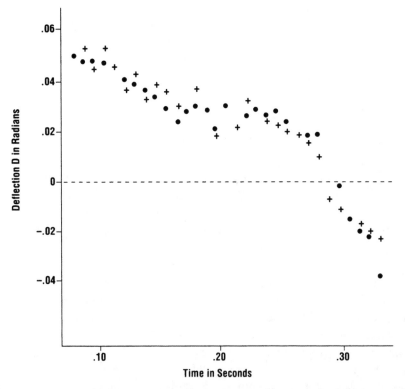

FIGURE 12.1. How the angle of deflection of the shaft in the plane of the swing in a particular swing by a professional golfer varied with time into the downswing. The crosses (+) and the dots (·) represent measurements of the deflection in two independent trials. The scatter of points is an indication of the difficulty in making these measurements. The general trend of the deflections in the downswing is quite obvious.

How Some Think It Flexes

One writer on this subject assumes that during the downswing the club oscillates as it does when clamped in a vice. The oscillatory process is described in the following manner. As the downswing begins, the inertial forces will first bend the club backward in the swing. In the terminology of this writer the shaft is said to "load." It acquires energy when it bends. Then during the downswing the shaft unloads, becomes straight, and loads again in the opposite sense, bending forward in the swing. As the swing progresses the shaft unloads, becomes straight, loads again bending backward in the swing, and unloads once more before the ball is hit. If you count the oscillations in this description of the swing, you will find that there are 1.5 of them, as discussed previously.

Any oscillating mechanical system, such as a simple pendulum, has its maximum speed at the midpoints of the oscillation and has zero speed at the maximum displacements of the oscillations. If this relation between displacement and speed were to hold in the swing of a golf club, the ball should be hit when the shaft of the club is straight, at which point the component of the speed coming from the oscillation of the shaft adds to the speed of the clubhead so as to send the ball on its way with a greater speed than would be the case for an inflexible shaft.

When we start to consider the effect of the flexibility of the shaft, we should perhaps consider the torque acting to bend the shaft resulting from the offset of the clubhead. The magnitude of this torque is given by the product of the mass of the clubhead, the length of the offset of the center of mass of the clubhead from the center of the shaft, and the centripetal acceleration of the clubhead. Early in the swing this torque contributes to the bending of the shaft toward smaller values of D but contributes little torque in this sense during the latter part of the swing. We shall neglect this torque.

The Vice Grip Calculation

The three differential equations were first solved for the case of a club with a flexible shaft, giving an oscillation frequency of 5.16 vibrations per second for the standard swing starting from rest at the start of the downswing. The oscillation frequency of 5.16 vibrations per second was measured for the driver clamped in a vice at the point where the golfer's thumb and first finger of his right hand would grip the club. The results of this calculation are shown in Fig. 12.2, curve A. In the figure the angular variable D, descriptive of the bending of the shaft, is plotted as positive when the bending of the shaft is backward shortly after the start of the downswing. We shall call this the vice grip flexible shaft calculation.

We see that the calculation describes the loading of the shaft in the backward direction between $t = 0$ and $t = 0.09\,\mathrm{s}$, the unloading between $t = 0.09$ and $t = 0.144\,\mathrm{s}$, the loading in the opposite sense (forward bending) between $t = 0.144$ and $t = 0.194\,\mathrm{s}$, an unloading until $t = 0.243\,\mathrm{s}$, and a final loading in the backward direction and unloading again by the time the ball is hit at $t = 0.334\,\mathrm{s}$. The calculated swing almost exactly fits the description of the swing of a club with a flexible shaft described earlier. The calculation shows an increase of 8.7% in clubhead speed at impact compared with that of the standard swing with an inflexible shaft. Figure 12.2, curve B, shows the time rate of change of the angular variable D for this calculation. We shall call it $\dot{D}$, since it is the angular speed of the oscillations of the variable D. It is plotted against time into the downswing. Notice the phase relation between the angular displacement variable D and the angular speed variable $\dot{D}$, shown in Fig. 12.2.

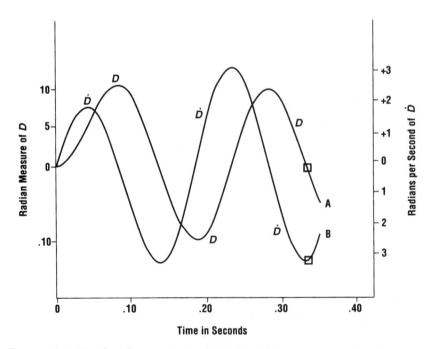

FIGURE 12.2. Results of a computer calculation of the standard swing when the club is gripped as it would be in a vice. Curve A shows the deflection that would occur in such a swing, in the plane of the swing as a function of time. Curve B shows the rate at which D changes with time.

Figure 12.3 shows two curves of the calculated clubhead speed plotted against time into the downswing. Curve A shows the speed calculated for the standard swing for an inflexible shaft. This curve fits the speed of the clubhead almost exactly for the swing of a professional golfer as determined by stroboscopic photography. Curve B shows the speed of the clubhead from the vice grip flexible shaft calculation discussed previously. While these curves show only two crossings, they cross three times very close to those times when curve B of Fig. 12.2 shows the speeds of the vibration of the shaft of the club to be small, $\dot{D} = 0$. Between these times the two curves show conspicuously different speeds. This difference is well outside any possible errors of measurement from the photographs and raises the question of the validity of the model used in the vice grip flexible shaft calculation of the speed of the clubhead.

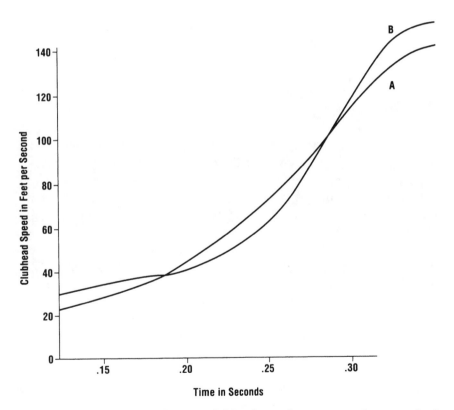

FIGURE 12.3. Two curves showing clubhead speed. Curve A is for a standard swing with an inflexible shaft. Curve B is for the vice grip flexible shaft calculation shown in Fig. 12.2. Curve B is completely outside the acceptable deviations from Curve A. Obviously, the vice grip swing is not what we find from experiment.

The Error in the Vice Grip Model

The error of our model is quite obvious. No golfer grips the club as it is gripped in a vice. In a second experiment, a club was clamped in a vice in the usual way but with a layer of sponge rubber surrounding the grip. The frequency of oscillation of the club when clamped in this manner was markedly lower than when clamped without the rubber. While the properties of a golfer's hands cannot be matched by those of sponge rubber, the lowering of the frequency seems to be a possible modification of the model. When one person holds a golf club by the grip and another pulls the clubhead to one side and releases it, the motion of the clubhead should give an indication of the frequency of the oscillation of the club during the downswing. In one case where a professional golfer held the club, when the club was released, the club-head had a transient translational motion but showed no sign of oscillation. In a second case, again with a professional golfer, the response was similar, with the least hint of a slow oscillation.

A Better Model

We are faced with finding a model that will give us a calculation for the bending of the shaft in terms of the angular parameter D that follows the general distribution of experimental points shown in Fig. 12.1. The parameters that must be chosen for the calculation should throw some light on what the golfer does in the downswing to influence the bending of the shaft.

If we extrapolate the trend of the experimental points in Fig. 12.1 to the time $t = 0$ at the start of the downswing, we find that the club is bent through an angle we shall call $D(0)$. Obviously, this extrapolation is uncertain. $D(0)$ seems to be about 3.7 deg. As the downswing progresses from $t = 0$ something happens through the action of the golfer's hands to gradually reduce the frequency of oscillation of the shaft from some initial value at $t = 0$ to some low value. The bent shaft relaxes somewhat during this time. This relaxation time was finally estimated to be 0.135 s. For a calculation with this change in the model the variable D followed the scatter of experimental points into the downswing to about $t = 0.17$ s. After this time the value of D falls much too abruptly, going through $D = 0$, the shaft being straight, at about $t = 0.25$ s instead of through $D = 0$ at about $t = 0.29$ s shown by the experimental points. Figure 12.4, curve A, shows these values of D.

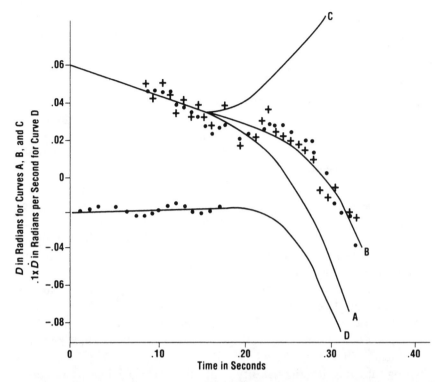

FIGURE 12.4. Three calculations made of the deflection angle of the shaft with various actions of the golfer's hands and wrists during the downswing. Curve A shows a calculation in which the golfer's hands exert a more realistic grip on the club compared to the vice grip. Curve B shows a calculation in which the golfer uses his wrists to exert a small helping torque in the wrist uncocking process. Curve C shows a calculation in which the golfer uses a larger uncocking torque. Curve D shows the calculated rate at which Curve B is changing.

Wrist Action and the Flexing of the Shaft

If in the latter part of the downswing the golfer uses his wrists to apply a torque to help in the uncocking of the wrists, the inertial force on the clubhead resulting from the acceleration produced by this torque will delay the unbending of the shaft. If we add such a torque by the wrists to our model, we may adjust the unbending of the shaft to fit what actually happens in the downswing. It turns out that a very small torque applied later in the downswing by the wrists to help in the uncocking process is sufficient to make the calculated value of D fit the experimental points as closely as they can be determined. The torque exerted by the wrists when a driver is held horizontally by the grip in the usual manner is about 2 lbs·ft. The torque needed in the calculation is only about 0.6 lbs ft, and it starts to be applied at $t = 0.14$ s, at which time the wrists begin to uncock naturally. Figure 12.4, curve B, shows the values of D obtained when this small torque is applied. We shall call this the relaxed grip flexible shaft calculation. Curve C in this figure shows the trend of the values of D when a torque of 2 lbs·ft is applied by the wrists at $t = 0.14$ s.

The calculations of the values of D also give the values of the rate of change of D, called $\dot{D}$, which is the speed at which D is changing. The values of $\dot{D}$ for Curve B of Fig. 12.3 are plotted in curve D in that figure. Notice that the phase relation between D and $\dot{D}$ here is not at all similar to that found in Fig. 12.2. For the bending of the shaft, as it is found in the stroboscopic photographs, the contribution of the bending motion to the clubhead speed is not at its greatest value when the shaft of the club becomes straight. This contribution to the clubhead speed continues to increase all through the swing until the ball is hit. It is definitely not at a maximum when the shaft is straight.

The values of the clubhead speed calculated from the quantities plotted in Fig. 12.4, curves B and D, when plotted against time, follow the curve of the standard swing at low speeds and deviate above it for higher speeds as the contribution of the flexing of the shaft becomes significant. There is nothing in the difference between these two curves calling into question the validity of the newer model. The gain in club-head speed at impact, coming from the contribution of the flexing of the shaft, is only about 3% according to these relaxed grip flexible shaft calculations. A similar calculation with the flexibility of the shaft increased by 10% produces essentially no change in this number.

There may be other explanations of the flexing of the shaft as found in our particular photograph of a swing of a professional golfer. However, when I confronted the gentleman who swung the club with my conclusion, that he must use an uncocking torque by his wrists when his wrists begin to uncock naturally, he confirmed that indeed he did apply such an uncocking torque.

Flex and Loft of a Club

The usual angle of loft of a driver is about 10 deg. The flexing of the club in the swing when the ball is hit is about 3.3 deg. This flex will increase the effective loft of the club. Certainly, the exact flex of such a club will depend on how the golfer uses his wrists during the down-swing. We have also seen that what the golfer does with his wrists also affects the wrist-cock angle at impact; a late hit will effectively decrease the angle above the horizon with which the ball will leave the clubface. We should not be surprised that a club that gets the ball up satisfactorily for one golfer may not do so for another with a different swing.

The data we have used in this study are from one swing of one golfer. I think that the flexing of the shaft forward as the clubhead comes in to hit the ball is a general characteristic of the golf swing. Photographs made with the use of a focal plane shutter must be ignored in this context. There are few photographs in the golf literature made with stroboscopic light sources, but of those I have examined, all of them show clubs flexed in this manner.

CHAPTER 13

Examining the Handicap System

Probability Theory

In the year 1654 two gentlemen who were playing at a game of chance were interrupted, and the game was not completed. They asked the famous French mathematician Blaise Pascal to determine from the state of the game when they left the table a fair distribution of their wagers. Pascal wrote a letter to another French mathematician, Pierre Fermat, telling him of this problem. Each solved it, though in different ways. The mathematical theory of probability developed from their work.

Blaise Pascal, 1623–1662, was educated by a tutor who was forbidden by Pascal's father to teach the boy any mathematics. At the age of 12, the boy asked his tutor what geometry was about. The tutor answered the question, and since it was a forbidden subject, Pascal had an incentive to spend his play time doing research in geometry. By the age of 14 he was admitted to the weekly meetings of a group of French geometricians.

Pierre de Fermat, 1601–1665, the son of a leather merchant, was educated at home. His interests were in geometry and the theory of numbers. These two gentlemen share the honor of creating the mathematical theory of probability.

While its origins were in games of chance, probability theory has been found to have many other practical applications. The immense insurance industry of the world is based on the expectations of individual policy holders. Industrial quality control and public opinion polls both involve sampling techniques based on probability theory. Probability theory continues to be applied in the study of games in which chance is a factor.

The fascination we find in many games comes from testing our skill against blind chance in determining a game's outcome. Many of us are not interested in games that depend mainly on skill or mainly on chance. It is the proper combination of the two that works to maintain our interest in a game. Golf is a game in which chance and skill are mixed in an interesting proportion, with chance playing a greater role than most players realize.

Two golfers of equal skill can compete on the course, making a golf

118

match a social event. When two golfers of unequal skill play together, each is playing against himself. The handicap system has been devised and continually improved to allow two golfers of unequal skill to play against each other on a more or less equal basis, although the chance aspect of the game still comes into play. The psychological response to winning a close match may be a feeling of "I did it!" interpreted in the realm of skill, but possibly a more rational response would be. "I was certainly lucky today!" interpreted in the realm of chance.

Tossing Coins: The Law of Large Numbers

We shall look at some of what probability theory can tell us about the handicap system. Before we do this we should look at a very simple application of this theory to elucidate the point of the discussion. Let us look at probability applied to tossing a coin. When we rule out the possibility of a coin standing on edge, we can say, from considerations of the shape of the coin, that heads and tails are equally likely to turn up when a coin is tossed in a random fashion. It is convenient to represent certainty by the number one. Since it is certain that we get either heads or tails on the toss of a coin, the probability of tossing a head or a tail is one-half. The sum of these equal probabilities is the probability of a sure thing.

There is another way to determine the probability of obtaining heads in the toss of a coin. We may proceed by tossing a coin a large number of times and counting the number of heads that turn up. The number of heads that turn up divided by the total number of coins tossed may be interpreted as the probability of tossing heads on any particular toss of a coin. In such a procedure the fractional variations of this ratio are expected to decrease as the number of tosses increases. We may say that it is highly probable that they do. The probability determined in this manner will likely not have exactly the same numerical value as that determined by considering the shape of the coin. In fact, it may be quite different if the coin is tossed only a few times.

Life insurance rates are determined by experience, which is reflected in the mortality tables. The probability of a person dying at a certain age can be determined in no other way. It may be of interest to look at the results of an attempt to determine the probability of getting heads on the toss of a coin by the second method. I tossed a coin many times, and I plotted the ratio of number of heads to the number of throws in Fig. 13.1 as the number of throws increased. There is a "law of large numbers" in probability theory that says it is highly probable that this ratio will approach the value one-half as the number of throws increases indefinitely. Of course, this ratio does not have to do this; it is entirely possible that a preponderance of heads or tails will occur. In

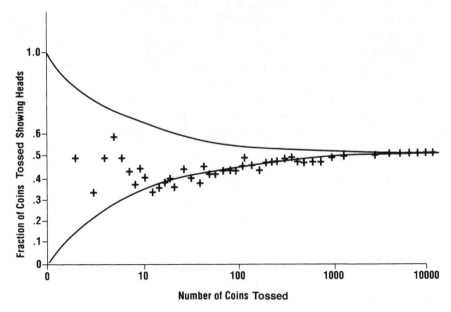

FIGURE 13.1. An experimental example of the determination of the ratio of heads to the number of coins tossed as 10,000 coins were tossed. The ratio approaches 0.5 for the large number of tosses, as is expected according to the law of large numbers. The smooth curves in the figure show the fractional standard deviation of the ratio from the value one-half as it depends on the number of coins tossed.

our demonstration the ratio was within 2% of the value one-half at 1000 tosses and within 0.22% at 10,000 tosses. The smooth curves in the figure show the fractional standard deviation of the ratio from the value one-half as it depends on the number of coins tossed.

Probability of More Complicated Events

Once the probability of the occurrence of a single event, such as the toss of a single coin to produce heads, has been determined, then the theory of probability may be used to determine the probability of more complicated events [23]. We shall consider the probabilities related to the toss of two coins as an example. In the toss of two coins, the sequencences HH, HT, TH, and TT are equally probable. Each occurrence thus has the probability of one-fourth. This probability is the product of the probability of the first coin showing heads or tails and the probability of the second showing heads or tails. We say that the probability of getting two heads (or two tails) is one-fourth, while the probability of getting heads and tails in either order is twice as great, one-half. These three probabilities add up to one because it is certain that one of these events will happen on the tossing of two coins. The reader may be interested in extending these ideas to the tossing of three or more coins.

The question, of course, remains as to whether these calculated probabilities have anything to do with what one finds when two coins are tossed. If they do, we should get heads and tails twice as often as we get two heads (or two tails). This is something that can be checked by tossing two coins at once. If one does this, the law of large numbers should be kept in mind. Even with only a few tosses of two coins the disparity of the probability of throwing two heads (or two tails) with that of tossing heads and tails may easily be demonstrated.

Probability Applied to Golf

When we turn to applying probability theory to scoring the game of golf, many golfers may object because they do not usually think of variations in their scores in terms of chance. They may maintain that scores are known to vary because of many factors such as the condition of the course, the weather, and the psychological atmosphere of the competition. They think of these as causal factors affecting the score. Consider the analogous application of probability theory in the life insurance industry. People die from many definite known causes. It is the influence of many independent factors on the human life span that allows the industry to be based on probability theory. In a similar fashion it is the influence of many independent factors that allow us to look at the scoring in the game of golf from the viewpoint of chance.

Chance most obviously enters in the action of the golfer in making the golf shot. First, he must draw on his past experience in judging what he should attempt. The factors involved here are almost infinitely varied. The individual stroke is such a complicated dynamic event that it is physically impossible to bring it off without uncontrollable variations from the stroke intended. Slight errors at impact become grossly magnified in the flight of the ball. Unknown conditions of the wind, the condition of the course where the ball lands, and the quality of the lie of the ball for the next shot all bring additional chance variations into the shot. The chancy nature of the play on the green is recognized by everyone. The pressure toward the end of a close match may be the result of a putt spun out of the cup earlier in the match when it might just as well have dropped.

The golfer tries his best to finish the round with the lowest possible score. Subjectively he feels that to some extent he has causal control over what the ball does. Objectively he knows that chance happenings have a great deal to do with the final outcome. It is this latter aspect of golf that can be illuminated by probability theory.

The Chance of Having a Really Good Round

It appears that every golfer lives with the hope that some day, even with his present skills, he will "put together" a really good round of golf. We may apply probability theory to determine the probability of a golfer with a particular level of skill of having a round with a particular score. To be definite, let us look at the chances of a particular golfer breaking 80 during a recent golfing season.

To make such a determination, the individual probabilities of making certain scores on each of the 18 holes of the course must be known. This golfer had kept his score cards for the rounds of golf he played during part of that season. From these score cards I found that on a particular par four hole he had shot 2 birdies, 3 pars, 14 bogies, 2 double bogies, and 1 triple bogie in 22 rounds of golf. From these numbers I calculated rough probabilities of him shooting particular scores on this hole. As an example, the probability of him having a par on this hole is 3 divided by 22, or 0.136. Obviously, he had not played enough to determine precise probabilities according to the law of large numbers. I must be content with these estimates of the probabilities. Using the probabilities determined for the various scores on each of the 18 holes, I was able to calculate the probability of him obtaining any particular score in a round of golf [23]. The calculation is so laborious that I called on a computer to do the work for me.

As I expected, the probability that he would put together a good round of golf, that is, break 80, turned out to be very small. The computer printout told me quite definitely that with the skill he had attained at that time, the probability that he would shoot a 79 (break 80) in a round of golf was less than 3 in one hundred thousand. In other words, if he were to play a round of golf every Saturday afternoon for 2000 years, he might break 80 by one point three times from purely chance variations in his scoring. If he had done his best on each hole for one round of golf, he would have had a score of 72, but the probability of such an occurrence is less than 1 in a million.

Probability Applied to the Handicap System

Except when we are polishing our own game, golf is usually a social activity. Golfers like to compete with one another. The handicap system has been devised to allow golfers of differing levels of skill to compete on an equal footing [24]. Probability theory may be used to study the handicap system to determine whether it does what it is purported to do. As we shall see, we are led to suspect that the handicap system does not quite put golfers of different levels of skill on an equal competitive basis. Some knowledgeable golfers say that it does not and should not. The handicap system should be biased slightly in favor of the more skillful golfer to give the less skillful golfer an incentive to improve his game.

For this discussion a golfer's handicap is to be established by taking 95% of the difference between the course rating and the average of the 10 best of his 20 latest 18-hole scores. This result is to be rounded off to the nearest whole number. In stroke play the net score is determined by subtracting the handicap from the actual number of strokes used in the round of golf. In match play the higher handicap golfer is given strokes on certain holes in the descending order of their difficulty. The total number of strokes given him is equal to the difference of the handicaps of the two competing golfers.

Comparison of the Scores of Two Golfers

In my study of the handicap system I have made use of score cards of two golfers who played together on a particular course. These gentlemen kindly supplied me with 20 consecutive score cards, from which I calculated probabilities for each of them getting a particular gross score on each hole. On a certain par four hole, for example, the better golfer had probabilities of 0.05 for a birdie, 0.40 for a par, 0.50 for a bogie, and 0.05 for a double bogie, while the poorer golfer had probabilities of 0.05 for a birdie, 0.20 for a par, 0.55 for a bogie, 0.10 for a double bogie, and 0.10 for a triple bogie. The poorer golfer had as much chance of shooting a birdie on this hole as the better golfer, but he was more likely to shoot double and triple bogies.

Using the individual probabilities thus determined on each of the 18 holes for both golfers, I set the computer to calculate the probability of each golfer shooting any particular gross score in a round of golf. The results of these calculations are shown in the two curves in Fig. 13.2. Although the probabilities are determined only for integer scores, I have drawn a smooth curve through the values given by the computer as an aid in seeing how they vary with the score for 18 holes. The most probable score for the better golfer is 79 and for the poorer golfer is 84.

When we come to look at the handicap system in terms of probability theory we run into one difficulty. The calculations of the probabilities relating to the scores of the two golfers were based on the assumption that the scoring depends on many varying factors, making it essentially a random process. When two golfers play together in a match, their scoring is influenced by the same conditions of the course and the same

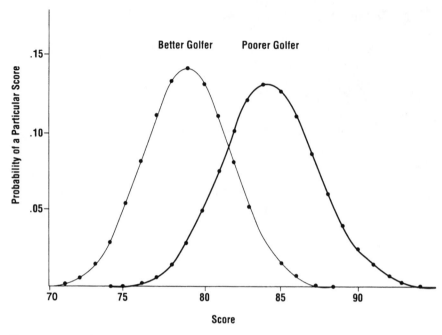

FIGURE 13.2. The probability of a particular gross score plotted against the score for each of two golfers.

aspects of the weather. Some of the randomness in the scoring is thus removed. We shall temporarily ignore this nonrandomness and consider later how it may influence our thinking concerning the handicap system.

The Better Golfer Has the Advantage

Let us consider what the application of the handicap strokes does to the two curves in Fig. 13.2. The better golfer and the poorer golfer were found to have handicaps of six and ten, respectively. In stroke play these handicap strokes are subtracted from the gross scores to obtain the net scores. The net scores so obtained are shown in the two curves in Fig. 13.3. These curves show that the calculated net scores with the highest probabilities differ by only one stroke, while the calculated gross scores with the highest probabilities, as shown in Fig. 13.2, differ by five strokes. The fact that these curves in Fig. 13.3 are not more nearly superposed suggests that with the handicap system used in this study applied to stroke play, the better golfer maintains a distinct advantage over the poorer golfer.

We may determine the advantage enjoyed by the better golfer by

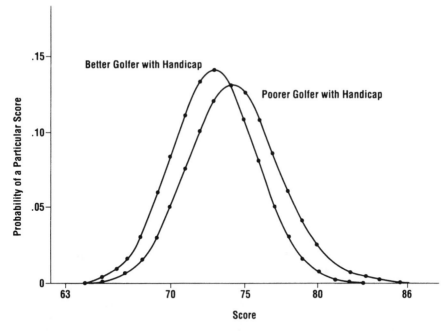

FIGURE 13.3. The probability of a particular net score at stroke play plotted against that score for two golfers after consideration of their handicaps is taken. Notice that the better golfer appears to have an advantage.

calculating the probabilities that he will win, tie, or lose a particular match at stroke play. This calculation shows that these probabilities are, respectively, 0.58, 0.09, and 0.33. According to the score cards, he actually won 60%, tied 15%, and lost 25% of the matches at stroke play using the handicaps mentioned before. The close agreement between the number of matches won, tied, and lost using these handicaps and the number predicted according to the calculated probabilities should give us confidence in the results of our probability calculations. Both the actual results and the calculated results indicate that indeed the better golfer, at least in this particular case, has a definite advantage over the poorer golfer.

Let us look at the question of whether the two golfers would be on a more even footing if the poorer golfer were allowed to increase his handicap by one stroke, from 10 to 11. From Fig. 13.3 we can see that moving the one curve closer to the other by one stroke would almost superpose the two curves. With the handicap difference of five strokes, instead of four strokes, the probability of the better golfer winning a match at stroke play is found to drop to 0.48. Thus he would probably not win half the matches with the new handicap for the poorer player.

Examination of the score cards showed that indeed he did not win half the matches using these handicaps.

Does this change of the handicap for the poorer golfer from 10 to 11 put the golfers on an even footing? If we mean by an even footing that the golfers have the same probabilities to win, tie, or lose a match at stroke play, then this extra handicap does not do the job. With the new handicap, the better golfer has probabilities of 0.48, 0.09, and 0.43 to win, tie, or lose, while the poorer golfer has probabilities of 0.43, 0.09, and 0.48 to win, tie, or lose. Giving the poorer golfer yet another extra stroke, raising his handicap to 12, would give him a definite advantage over the better golfer. The score cards show he would actually win 65% of the matches with such a handicap at stroke play.

In match play with a handicap difference of four strokes the poorer golfer will be given one stroke on each of four designated holes. The granting of a stroke on any particular hole changes the probabilities for scores on that hole for the poorer golfer. The probability for scoring a par on the hole becomes, for the purpose of the calculation, the probability of scoring a birdie. There is a similar adjustment on the scores for all the other probabilities on the hole. A calculation was made, using probabilities adjusted in the manner just described, to find the probabilities that the better golfer would tie, win, or lose by various numbers of holes in match play. The results are shown plotted in Fig. 13.4. The zero on the scale along the base represents the tie score in a match. To the right the scale indicates the score by which the better golfer might win a match and to the left the score by which he might lose a match. The height of the curve represents the probability of the better golfer winning, tieing, or losing by the corresponding score. As an example, the probability of the better golfer winning a match by one hole is 0.157. The bell shape of the curve indicates that when the handicap difference is taken into account neither golfer will often win or lose by a large amount.

The distribution of probabilities in Fig. 13.4 is not symmetric about zero on the score scale. The probabilities of the better golfer winning or losing a match are not equal. An additional calculation showed that at match play, with a handicap difference of four, the probability of the better golfer winning is 0.58, which is also his probability of winning at stroke play with that handicap difference. The better golfer thus has essentially the same advantage in both match play and stroke play. According to the score cards, the number of matches won by the better golfer was the same at match and stroke play. However, four matches scored differently depending on whether they were evaluated according to match play or stroke play. For instance, one match was won at match play and tied at stroke play, while another was tied at match play and won at stroke play.

If the poorer golfer was allowed an extra handicap stroke, giving him

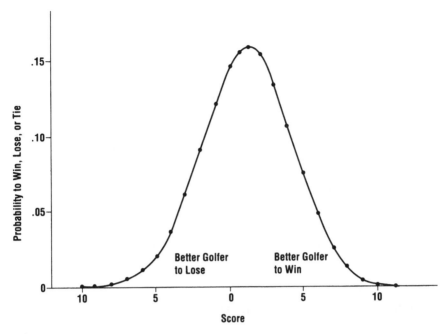

FIGURE 13.4. The probability of the better golfer to win, lose, or tie at match play plotted against the score. Scores to win are plotted to the right of zero on this scale and those to lose are plotted to the left of zero. Notice that at match play the better golfer appears to have an advantage.

a handicap of 11, the result of a similar calculation for a handicap difference of five strokes showed that the better golfer had a probability of winning of 0.50, very slightly higher than for stroke play with the same handicap difference. Even this increase in handicap does not put the two golfers on an even footing, since the probabilities for the better golfer to win, tie, and lose are now 0.50, 0.13, and 0.37, and the probabilities for the poorer golfer to win, tie, and lose are 0.37, 0.13, and 0.50. Granting the poorer golfer an additional handicap stroke, which would increase the handicap difference to six strokes, gave the poorer golfer a distinct advantage over the better golfer.

Rounding Off to the Nearest Whole Number

There is an aspect of the application of the handicap system that should be mentioned so that a possible misconception may be avoided. In calculating handicaps the last step is to round off to the nearest whole number. For example, when the handicap calculation gives 5.49, the handicap is rounded down to five strokes, but if the calculation had given 5.51, the handicap would have been rounded up to six strokes. The change from a handicap of five to six could result from a single missed putt by the golfer in one of his ten better matches. In the examples for the two golfers with handicaps of six and ten strokes, both handicaps were rounded off to lower numbers. Giving the poorer golfer an additional stroke, raising his handicap to 11 strokes, would correspond to a case where his handicap calculation was rounded off to the higher number. The handicap difference would be raised to five strokes. A corresponding rounding off of the better golfer's handicap calculation to the higher number would give him a handicap of seven strokes. The handicap difference would then be lowered to three strokes. Such a swing in handicap difference from five to three makes the probability of the better golfer winning or tieing (not losing) at match play vary from 0.63 to 0.78. A similar variation would occur at stroke play. In no case could this variation in handicap determination put the two golfers on an even footing.

Is Scoring at Golf a Random Process?

The calculated probabilities presented in this discussion are based on the assumption that scoring at golf is a random process. If the better golfer should turn in a score obtained on a given course under a particular set of conditions of the course and weather and if the poorer golfer should turn in a score obtained on the same course but under conditions and weather unrelated to those found by the better golfer, then our calculations would give valid probabilities for one of the golfers to have a higher, equal, or a lower net score than the other. If our golfers go out on the course at the same time, as they would if they were playing a match, then the assumption mentioned previously does not apply. The scoring in such a match no longer can be considered to be a completely random process.

In the estimation of the previously indicated probabilities, the contribution to the probability that the higher handicap golfer wins the match comes mainly from scores that are exceptionally low for the higher handicap player and exceptionally high for the lower handicap player. Under the condition of randomness this particular arrangement of scoring is likely to occur when the poorer golfer plays under exceptionally good conditions and the better golfer plays under very adverse conditions. When they play under the same conditions, it is likely that any influence on the scoring would tend to be similar for both golfers. However, since the better golfer is more able to take advantage of good conditions and less likely to be adversely affected by trying conditions, the probability of the better golfer winning a match would be higher than the probability calculated on the assumption of random scoring.

In other words, the nonrandomness introduced when two golfers play together increases the advantage of the better golfer above that previously shown. It is impossible even to estimate this increase in the advantage of the better golfer, but the score cards show that he actually lost only 25% of his matches at both stroke and match play, while the probabilities calculated for losing at stroke and match play are 0.30 and 0.33, respectively. The differences between actualities and probabilities are in the direction expected from the decrease in randomness when the golfers play together.

This analysis of the handicap system in terms of the experience of two golfers certainly raises doubts that it puts golfers of different skills on an equal footing. There is no doubt that it goes part of the way toward this goal, but no golfer should play under the illusion that it reaches this goal.

Changes in the Handicap System

This discussion has been concerned with handicaps determined for two golfers playing together on a particular golf course. When two golfers are competing on a course with handicaps determined by play on two other courses, then the value of the handicap system becomes very uncertain.

The problem of the "portability" of handicaps has been investigated, and a modification of the system, called the slope handicap system, has been tested and is now in use at over 10,000 golf courses in the United States. The modification has been discussed in the literature [3], and further description of this modification would be out of place here.

CHAPTER 14

A Short Chapter on Short Putts

Babying a Putt into the Cup

A golfer who is doing well in a national championship tournament standing over a crucial two-foot putt must experience very intense stress. In this circumstance a single putt may be worth $50,000 or more. If a person watch tournaments on television before long he will see a tournament lost or won depending on what a golfer does with a short putt. Failure to sink such a putt is reported as caused by nervous strain of the moment. This may be true, but the cause may also be the lack of the fundamental understanding and application of Newtonian dynamics. When you see a two-foot putt just barely roll past the hole, you should recognize that the golfer was trying to "baby" the ball into the cup.

Dave Pelz's Book on Putting

A very complete analysis of the art of putting has been carried out by Dave Pelz, a physicist, and reported in his book *Putt Like the Pros* [21]. Although he writes about the science of putting, there is little in his book about applying the science of mechanics to the art of putting. However, his understanding of putting is based on a careful experimental study of the field, and his book should be studied in detail by everyone with ambitions in golf. If Pelz has such an excellent book on the art of putting, is there more on the subject to discuss? There is. First, I wish to call the reader's attention to Dave Pelz's book. And second, I have found that physics has something to say about the importance of one of Dave Pelz's discoveries about short putts. He discovered that during the putting stroke, the ball in a missed putt should continue to roll seventeen inches past the hole. He gives no physical explanation of why this maneuver is important.

130

The Physical Theory of Short Putts

Newton's second law states that an unbalanced force acting on a body produces a rate of change of the momentum of the body. And thus a force acting on a body for a certain time produces a certain change of momentum of the body. This change in momentum may be a change in the speed of the body or it may be a change in the direction of the motion of the body, or both can happen.

Let us consider two putts on a sloping green with an uncertain surface to a cup two feet away. The first will be a putt that just reaches the cup; if it misses the cup, the ball will stop after going just past the cup. The golfer "babys" this putt. The second putt will travel two feet past the hole, the putt starting at the same position as the first and traveling along the same path. Of course, the golfer will have to hit the second putt harder than the first to have the ball go the four feet. For the first two feet of the second putt the ball will be moving more rapidly than the first ball over the same path, and the second ball therefore covers the first two feet in less time than the first ball.

If we assume that the retarding forces between the balls and the surface of the green is not significantly different over the first two feet as a result of the differences of ball speeds, the total change of momentum of the more rapidly moving ball will be less than for the more slowly moving ball. The second putt will thus have a smaller deflection of the ball over the first two feet and will have less chance of missing the cup.

On every putt longer than two feet, there is a last two-foot section. This means that the conclusion we have found for a two-foot putt holds for every putt greater than two feet in length. Dave Pelz could have used a two-foot roll-by as well as a seventeen inch roll-by in his discussion. We find that there is a firm physical basis for his discovery. He did not arrive at his discovery by the application of theory, but rather by an extensive experimental investigation. Understanding the theoretical basis for this result should tell the golfer that indeed such a roll-by should occur on every missed putt; this result should receive serious consideration by every golfer.

A Short Backswing on Putts

The backswing on a putt is an area for research that every aspiring golfer should find of interest. May I suggest that the golfer find a level area of short napped carpet. Use two golf balls, placing them two feet apart. Putt one ball so that it hits the other. The collision of the two balls will tell the investigator whether his putt is true to the mark. If the two balls have a perfect collision, the putted ball will stop, and the other will roll in the direction of the putted ball. If the putted ball misses a head-on collision ever so slightly, then the two balls will separate, with neither ball going in the original direction of the putted ball. Thus, observing the motion of the two balls after the collision will show the quality of the putt in a dramatic fashion. The action of the two balls follows directly from the principle of the conservation of momentum.

In your study, first swing a putter farther back than is comfortable. Watch where the clubhead goes on each of these trials. I think that you will find that the farther back you swing the putter, the more trouble you will have in moving the clubhead along the correct line of the putt. Conversely, making the backswing shorter and shorter will likely give you greater precision in your backward swing and in your forward swing. Soon it may dawn on you: "Why have a backswing at all in strokes of short putts where you need the utmost in precision?"

The rules of golf clearly state that "the ball shall be fairly struck at with the head of the club and must not be pushed, scraped, or spooned." I cannot find anything in the rules concerning the length of the backswing. If the start of the swing is with the clubhead at rest a few inches behind the ball, the ball should be considered fairly struck. I find that for me the precision of such putts is much greater than with a back swing. A golfer may check this observation by putting on a level carpeted floor.

If this type of putt works for a golfer, then he may find that on certain putts, where the green has considerable slope and a rough, uncertain surface, he may be able to negate these conditions by hitting the ball firmly. Such a stroke will allow little time for the momentum to change laterally, and in this condition the speed of the ball is of little consequence. He may thus trust his putting and hit the ball smartly into the hole. The study of this method for relatively short putts should be an interesting research problem.

I was once invited to play a round of golf in the Seattle area with a very competent golfer. I watched while he missed some short putts that he should not have tried to "baby" into the hole. After the round was over, I showed him what I could do. He found that he could do likewise. He thanked me for the short lesson. I continued with, "That will be a thousand dollars, please," and we both laughed. I am sure that this short lesson would have increased the winnings one hundred times that amount for a certain individual I watched lose a national tournament by missing a couple of two-foot putts he tried to "baby" into the hole.

Why not work on your short putts as a research project. Perhaps you will be successful with this?

References

1. John Stuart Martin, *The Curious History of the Golf Ball* (Horizon, New York, 1968).
2. George Peper, *Golf in America* (Harry M. Abrams, New York, 1988).
3. *Science and Golf*, edited by A. J. Cochran (Chapman and Hall, New York, 1990).
4. Harold E. Edgerton and J. R. Killian, Jr., *Flash* (Charles T. Bradford, Boston, 1954).
5. Lloyd Goodrich, *Winslow Homer* (George Braziller, 1959).
6. B. Bernstein, D. A. Hall, and H. M. Trent, "On the dynamics of a bull whip," J. Acoust. Soc. Am. **30**, 1112 (1958).
7. Alistair Cochran and John Stobbs, *The Search for the Perfect Swing* (J. B. Lippincott, 1968).
8. Robert Tyre Jones, *Bobby Jones on Golf* (Doubleday, New York, 1966).
9. Jack Nicklaus, *The Greatest Game of All; My Life in Golf* (Simon and Schuster, New York, 1969).
10. Sam Snead, *The Driver Book* (Cornerstone Library, New York, 1967).
11. Tommy Armour, *How to Play Your Best Golf All the Time* (Simon and Schuster, New York, 1953).
12. Cary Middlecoff, *Master Guide to Golf* (Prentice Hall, Englewood Cliffs, NJ, 1960).
13. Frank Beard, *Golfer's Digest*, 1968.
14. Al Geiberger, *Golf*, Sept. 1972.
15. Julius Boros, *How to Win at Weekend Golf* (Fawcett, New York, 1964).
16. John Le Masurier, *Discus Throwing* (Amateur Athletic Assn., London).
17. Louis T. Stanley, *Master Golfers in Action* (Macdonald, London, 1956).
18. C. G. Knott, *Life and Scientific Work of P. G. Tait* (Cambridge University, London, 1911).
19. *Handbuch der Experimentalphysik*, 1 Teil (Akademische Verlagsgesellschaft mbh, Leipzig, 1931), Vol. IV, pp. 38–39.
20. Ancher H. Shapiro, *Shape and Flow*, Science Study Ser. S21 (Doubleday Anchor, New York).
21. John Joseph Thomson, Nature **85**, 251 (1910).
22. John M. Davis, "The aerodynamics of golf balls," J. Appl. Phys. **20**, 821 (1949).
23. T. P. Jorgensen, "On probability generating functions," Am. J. Phys. **16**, 285 (1948).
24. *The P.G.A. Book of Golf* (1968), p. 80.
25. Michael McDonnell, *Golf, The Great Ones* (Drake, New York, 1973).
26. R. D. Mahta, Ann. Rev. Fluid Mech. **17**, 151–189 (1985).
27. Dave Pelz, *Putt Like The Pros* (Harper and Row, New York, 1989).

TECHNICAL APPENDIX— SECTION 1

Parameters of Golf Clubs

Any scientific discussion of the characteristics of golf clubs must be based on their three dynamic parameters: the mass of the club, the first moment about some axis, and the second moment about the same axis. When a golfer takes a club in hand, he finds that his wrists cock about an axis close to 5 in from the grip end of the club. In all our discussions this distance will be arbitrarily taken to be 5 in.

Consider a golf club as shown in the following diagram.

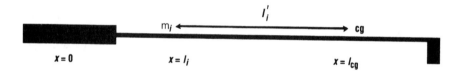

An origin of coordinates is chosen at the point on the shaft about which the golfer cocks his wrists. This point is at $x = 0$ in the diagram. In our imagination we are going to cut up the club into a large number of small pieces and weigh each piece, thus obtaining the mass m_i of each piece. The letter i is an index number indicating a particular small piece of the club. One such particular piece is located at $x = l_i$, or at a distance l_i from the axis about which the golfer cocks his wrists.

One dynamic parameter is simply the total mass M of the club, which is the sum of all the m_i and may be written

$$M = \sum_i m_i,$$

where Σ indicates the sum over all the values of i.

The second parameter is called the first moment about the wrist-cock axis and may be defined by the expression

$$S = \sum_i l_i m_i .$$

This expression means that we are to obtain the product of m_i and l_i for each small piece of the club and then sum these products for all values of i to obtain the quantity S. There is a very easy way to determine S.

Any body balances in Earth's gravitational field about any axis through a single point in the body called the center of gravity. The location of the center of gravity of a golf club may be found by simply balancing the club across a finger. If m_i is at a distance l_i' from the center of gravity of the club, then the torque produced on m_i about the center of gravity is proportional to the mass m_i multiplied by the distance l_i' . Since the sum of torques on the balanced club add to zero, we may write $\Sigma m_i l_i' = 0$.

From the diagram we observe that if l_{cg} is the distance from the wrist-cock axis to the center of gravity of the club, then $l_{cg} = l_i + l_i'$, or $l_i = l_{cg} - l_i'$. Since $S = \Sigma m_i l_i$, we have $S = \Sigma m_i (l_{cg} - l_i') = \Sigma m_i l_{cg} - \Sigma m_i l_i'$, which is equal to $l_{cg} \Sigma m_i$, since $\Sigma m_i l_i' = 0$. And since $\Sigma m_i = M$, the mass of the club, we find that we may obtain S by simply multiplying the mass of the club by the distance to the center of gravity from the wrist-cock axis. Thus

$$S = M l_{cg} .$$

The third parameter is called the second moment about the wrist-cock axis. Some call this parameter the moment of inertia. It may be defined by the expression

$$I = \sum_i l_i^2 m_i .$$

When a body is allowed to swing about some axis as a pendulum, simple theory shows that for small oscillations the period T of the pendulum is given by

$$T = 2\pi \sqrt{\frac{I}{gS}}$$

where g is the acceleration of gravity. When T and S for the pendulum about some axis are known, then I about the same axis is found to be

$$I = \frac{T^2}{4\pi^2} gS.$$

Obtaining the second moment of a golf club then involves swinging it as a pendulum about the wrist-cock axis and determining its period of oscillation.

TECHNICAL APPENDIX— SECTION 2

The Matching of Clubs

The purpose of this section is to give the technical basis for the matching of a set of clubs. One club to which the other clubs of the set are matched we shall call the "master club." This club should be chosen with care, because the other clubs of the set will swing exactly the same as this club when the matching is perfect.

If a seven iron is chosen as the master club, such a club with a shaft of the proper stiffness may be modified in some way so that the club-head mass may be adjusted on the practice tee to have swing characteristics just right for the golfer for whom the matched set of clubs is to be produced. When the golfer has found the clubhead mass that makes this club just right for him, the mass and the first and second moments of this club about the wrist-cock axis are determined in the usual way.

It will be assumed that we have assembled a set of clubs to be matched to the master club. If club parts are at hand rather than the assembled clubs, the parts may be temporarily assembled. In either case, the dynamic parameters of the assembled clubs are also determined in the usual way.

In writing equations for the theory of matching clubs it is necessary to have a notation descriptive of the various clubs involved. The letters M, S, and I will be used to indicate the mass, the first moment, and the second moment, respectively, of a club. The letter B will be used to indicate any parameter of any club before any modification is made to the club. The letter A will be used to indicate any parameter of any club after any modification is made in a club to match it to the master club. The letter i will designate any club of the set other than the master club; the letter j will be used to designate the master club. Two examples will illustrate the notation: $IB(j)$ will indicate the second moment of the master club; $SA(i)$ will indicate the first moment of the ith club after it is modified to match the master club.

In the matching process, clubs may be modified by adding mass $A(i)$ to the head of the ith club and distributing mass $ms(i)$ along the shaft of the club. The mass distributed along the shaft will be designated by the expression

$$ms(i) = h \int f(x)dx,$$

where x is a coordinate along the shaft with the origin at the wrist-cock axis and where $hf(x)$ is the linear mass density of the distribution. Here h is a constant, and $f(x)$ is a function of x, continuous or discontinuous, yet to be determined. The integral extends over the length of the shaft.

The mass $MA(i)$, the first moment $SA(i)$, and the second moment $IA(i)$, of the ith club after modification will be

$$MA(i) = MB(i) + A(i) + ms(i) + mg(i), \tag{1}$$

$$SA(i) = SB(i) + A(i)L(i) + h \int xf(x)dx, \tag{2}$$

$$IA(i) = IB(i) + A(i)L(i)^2 + h \int x^2 f(x)dx. \tag{3}$$

Here $mg(i)$ is a mass placed at $x = 0$, in the shaft at the wrist-cock axis, so that this mass will have no effect on either $SA(i)$ or $IA(i)$. $L(i)$ is the coordinate of the center of mass of the clubhead. The terms $A(i)L(i)$ and $A(i)L(i)^2$ are the contributions to the first and second moments of the club of any increase in the clubhead mass, and $h \int xf(x)dx$ and $h \int x^2 f(x)dx$ are the contributions to the first and second moments of any mass distributed along the shaft of the club.

For the clubs to be matched in the new sense, the following relations must hold:

$$IA(i) = (1 + p)IB(j),$$
$$SA(i) = (1 + q)SB(j),$$
$$MA(i) = (1 + r)MB(j). \tag{4}$$

These equations state that the dynamic parameters of the set of clubs will be matched exactly to those of the master club or to any predetermined variation from those of the master club. The quantities p, q, and r are the fractional variations in the respective parameters of the set of clubs from those of the master club. The quantities p, q, and r may be positive or negative. When $p = -q = 0.005$, $IA(i)$ will be 0.5% larger than $IB(j)$, and $SA(i)$ will be 0.5% smaller than $SB(j)$. For exact matching, p, q, and r are set equal to zero.

With these values of $IA(i)$ and $SA(i)$, equations (2) and (3) may be used to eliminate $A(i)$, with the following result:

$$h \int [L(i)x - x^2]f(x)dx = L(i)[SA(i) - SB(i)] - [IA(i) - IB(i)]$$

$$= K(i). \tag{5}$$

Notice that each of these expressions is a constant. The mass distributed along the shaft $ms(i)$ becomes

$$ms(i) = h \int f(x)dx = \frac{K(i) \int f(x)dx}{\int [L(i)x - x^2]f(x)dx}. \tag{6}$$

This equation may be put into the form

$$\int [L(i)x - x^2 - K(i)/ms(i)]f(x)dx = 0. \tag{7}$$

This integral is of considerable interest because it can be shown that for a given $K(i)$, the minimum value of $ms(i)$ needed to match two clubs is

$$ms(i) = 4K(i)/L(i)^2, \tag{8}$$

and the mass must be placed at $x = 0.5L(i)$. The function $f(x)$ for this minimum value of $m(i)$ is the Dirac delta function $\delta[L(i)/2 - x]$. Other functions $f(x)$ may be used to match clubs, but the mass $ms(i)$ distributed along the club will be larger than $4K(i)/L(i)^2$. When $ms(i)$ is the minimum value, $A(i)$ is found to be

$$A(i) = 2[IA(i) - IB(i)]/L(i)^2 - [SA(i) - SB(i)]/L(i), \tag{9}$$

and $ms(i)$ is found to be

$$ms(i) = 4\{[SA(i) - SB(i)]/L(i)\} - \{[IA(i) - IB(i)]/L(i)^2\}. \tag{10}$$

In general, if $i < j$, $ms(i)$ will be positive and $A(i)$ will most likely be negative.

When the quantities $IB(j)$, $SB(j)$, $L(i)$, $IB(i)$, $SB(i)$, p, and q are known, these equations may be used to calculate $IA(i)$, $SA(i)$, $A(i)$, and $ms(i)$. Also, with the quantities $MB(j)$ and $MB(i)$, the sum $MB(i) + A(i) + ms(i)$ may be found. This sum together with $mg(i)$ may be used to calculate $MA(i)$ according to equation (1). The quantity $mg(i)$ may be chosen to bring $MA(i)$ to any desired values without affecting the already determined first and second moments of the clubs.

TECHNICAL APPENDIX— SECTION 3

Clubhead and Ball

When a golfer swings a club, he is preparing for a collision between the clubhead of equivalent mass M having a velocity V and a stationary ball of mass m. While the details of the collision, which lasts about 0.0005 s, are seemingly impossible to measure or calculate, the application of physical principles gives the golfer a feeling for some of the things that happen in this very short time.

Just before the collision, the orientation of the clubhead may be specified by a vector N, the normal to the face of the club, at an angle θ to the velocity vector. The two vectors V and N define a plane, which will be called the D plane, since it is descriptive of the collision. This plane is not necessarily a vertical plane. Unless the hands of the golfer are neither ahead nor behind the clubhead at impact, the angle θ will not be the loft of the club.

Just after the collision the clubhead will be moving with a velocity W, smaller than V, but essentially in the same direction as V. It is kept moving in this direction by the force applied by the golfer along the shaft of the club. The ball is also moving in the D plane, since during the collision, no impulse is applied to the ball perpendicular to this plane. The velocity vector U of the ball will lie between the vectors W and N at an angle φ to W. The ball will be spinning about an axis perpendicular to the D plane in a sense that will produce lift on the ball. The lift vector will also lie in the D plane. These various vectors are shown in Fig. A3.1.

Since the clubhead is supported on a long flexible shaft, little if any impulse along the path of the clubhead can be supplied by the golfer during the short time the clubhead is in contact with the ball. To a very good approximation the momentum along the path of the clubhead is unchanged during the collision.

The momentum vector just before the collision is simply the product of the clubhead equivalent mass M and the velocity of the clubhead V and is in the direction of the velocity vector V. The momentum in this direction after the collision, to the approximation mentioned above, consists of the momentum vector MW and the component of the

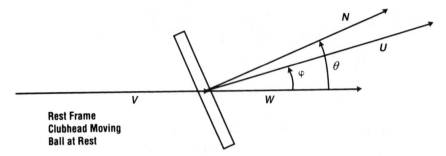

FIGURE A3.1. The vectors describing the clubhead–ball collision in the frame at rest relative to the ball before impact. N represents the normal to the clubhead surface, V the clubhead velocity before the impact, W the clubhead velocity after impact, and U the velocity of the ball as it leaves the clubhead. Theta (θ) is the angle between W and N, and phi (φ) is the angle between W and U.

momentum vector of the ball, $mU \cos \varphi$. The momentum of the club and ball taken together along this direction is unchanged during the collision. Therefore, we may write

$$MV = MW + mU \cos \varphi. \qquad (1)$$

In arriving at this equation we have looked at the collision and all these velocities in a frame of reference at rest relative to the position of the ball before the collision. We shall call this frame of reference the "rest frame."

In the further discussion of the collision we shall need to consider velocities in the frame of reference at rest relative to the clubhead. We shall call this frame of reference the "moving frame." Before the collision, the moving frame is moving to the right in Fig. A3.2 with a velocity V relative to the rest frame, and after the collision the moving

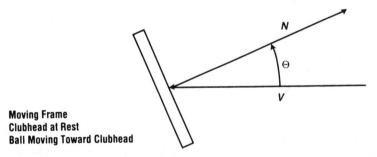

FIGURE A3.2. Two vectors in the moving frame, the frame in which the clubhead is at rest. In this frame the ball approaches the clubface with a velocity V, and N is the normal to the clubface.

frame is moving to the right with a velocity W relative to the rest frame.

When an elastic ball is dropped vertically along the normal to a smooth, hard floor, it will rebound to a height somewhat less than that from which it was dropped. Energy is lost in the compression of the ball. The collision between the ball and the floor can be described using an empirical relation called Newton's rule. Over quite a range of velocities, the ratio of the velocity with which the ball rebounds from the floor to the velocity with which the ball approaches the floor is approximately constant. This constant, e, is called the coefficient of restitution. If the ball hits the floor in a direction other than along the normal to the floor, the coefficient of restitution is approximately equal to the ratio of the normal components of the velocities previously mentioned.

Newton's rule may be applied to the collision between a golf ball and the clubhead surface when we look at the collision in the moving frame of reference. In this frame the ball approaches the clubface at an angle θ to the normal with a velocity V. The component of this velocity normal to the surface is $V \cos \theta$. The ball rebounds from the clubface with a velocity U in the rest frame at an angle φ to the velocity vector V. To obtain the normal component of the velocity of the ball in the moving frame, the velocity component $W \cos \theta$ must be subtracted from the normal component of U to obtain $U \cos(\theta - \varphi) - W \cos \theta$. For this collision Newton's rule takes the form

$$e = \frac{U \cos(\theta - \varphi) - W \cos \theta}{V \cos \theta}. \tag{2}$$

The velocities V, W, and U are rest frame velocities.

If we let $m/M = a$, we may solve equations (1) and (2) for W and U. We obtain

$$W = V \frac{\cos(\theta - \varphi) - ae \cos \theta \cos \varphi}{\cos(\theta - \varphi) + a \cos \theta \cos \varphi} \tag{3}$$

and

$$U = V \frac{(1 + e)\cos \theta}{\cos(\theta - \varphi) + a \cos \theta \cos \varphi}. \tag{4}$$

These equations tell us only part of what we want to know. For given values of V and θ, W and U can be found only when we know the angle φ. The value of this angle is in general very difficult to estimate with precision.

To lead into this problem it is of interest to look at a special kind of collision, one in which a perfectly elastic ball ($e = 1$) is hit with a very heavy frictionless clubhead ($a = 0$). For this kind of collision we derive

$$W = V \quad \text{and} \quad U = \frac{2 V \cos \theta}{\cos(\theta - \varphi)} \tag{5}$$

from equations (3) and (4). A diagram of the velocities involved is shown in the rest frame in Fig. A3.3.

The velocities we shall consider first are all in the moving frame of reference. In this diagram the ball approaches the clubface from the right along the horizontal with a velocity vector V at an angle θ to the normal N to the clubface. This velocity vector may be resolved into two components, a velocity $V \cos \theta$ toward the clubface and a velocity $V \sin \theta$ in an upward direction along the clubface. Since there is no friction, during the collision the ball slides up along the clubface with a constant speed $V \sin \theta$; the component of the velocity of the ball in this direction is the same after the collision as before the collision. Since with no friction and perfect elasticity no energy is lost during the collision, the component of the rebound velocity of the ball normal to the clubface is $V \cos \theta$. After the collision, the ball will be moving with a velocity V at an angle θ above the normal N to the clubface, since the

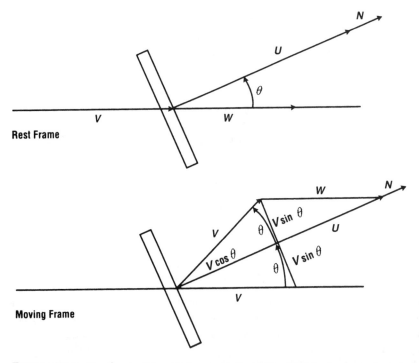

FIGURE A3.3. Similar to Fig. A3.1 except that Fig. A3.3 is for a perfect elastic ball and a very massive clubhead. In this case the vectors U and N are in the same direction. The upper figure is drawn for the rest frame, and the lower figure is drawn for the moving frame.

velocity vector is the sum of the two components $V \cos \theta$ and $V \sin \theta$. In this special case the ball rebounds with unchanged speed at the same angle on the other side of the normal N to the clubface.

To obtain the ball velocity in the rest frame of reference we must add the velocity W of the rest frame after the collision to the velocity of the ball in the moving frame. Since in this special case, as may be seen from equation (5), the velocity vectors V and W are equal in magnitude, these vectors add to give a velocity $U = 2V \cos \theta$ in the rest frame along the normal N to the clubface. This result is consistent with equation (5) when angles θ and φ are equal. When the angle θ is zero, the ball in this special case will have velocity $2V$ in the rest frame.

We are ready to consider a similar but more realistic collision between the ball and the clubhead in which there is friction between the ball and the clubface. We shall assume realistic values of e and a; $e = 0.7$ and $a = 0.19$.

In the moving frame the ball again approaches the clubface with a velocity V at an angle θ to the normal N. As shown before, the normal component of the velocity of the rebounding ball in the moving frame is $U \cos(\theta - \varphi) - W \cos \theta$. This quantity is equal to $eV \cos \theta$ according to equation (2). In the moving frame the ball is moving upward along the surface of the clubface. If it slides without friction it continues at velocity $V \sin \theta$, but if the force of friction is large enough, it starts to roll and by the end of the collision may be rolling upward along the clubface without sliding and with a smaller velocity.

It may be instructive to look at what happens when a bowling ball is thrown down a bowling alley. The ball starts to slide, and the force of sliding friction gradually sets the ball to rolling down the alley. By the time the ball hits the pins it has essentially pure rolling motion.

We may calculate the final speed of the bowling ball by equating the impulse on the ball from the possibly varying frictional force $F(t)$ opposing the motion of the ball to the decrease in the momentum of the ball along the alley. We may therefore write

$$\int F(t)\,dt = mv_1 - mv_2, \tag{6}$$

where v_1 is the initial velocity of the bowling ball, v_2 is it final velocity, and m is the mass of the ball. When the ball is rolling, $v_2 = R\omega$, where R is the radius of the ball and ω is the angular velocity of the rolling motion. We assume here that the radius of the ball remains essentially constant throughout.

Besides the linear impulse $\int F(t)\,dt$, a torque impulse $\int F(t)R\,dt$ also acts on the ball. This impulse torque is equal to the angular momentum acquired by the ball as it reaches its final velocity v_2. We therefore may write as a good approximation for a bowling ball

$$\int F(t)Rdt = 0.4mR^2\omega = 0.4mRv_2. \tag{7}$$

Here $0.4mR^2$ is the moment of inertia (second moment) of the ball about an axis through its center. If we regard R as constant, we may combine equations (6) and (7) to obtain

$$\int F(t)Rdt = R\int F(t)dt = Rm(v_1 - v_2) = 0.4mRv_2 \tag{8}$$

and thus

$$v_2 = \tfrac{5}{7}v_1. \tag{9}$$

The velocity v_2 is less than v_1 because the force of friction acting for the distance the ball slides removes energy from the ball and does work in setting the ball into rotational motion.

We may use this last result to estimate the speed of the golf ball as it rolls up the clubface at the end of the collision when we assume that the golf ball does not flatten against the clubface during the collision. Since the original speed of the ball up the clubface is $V\sin\theta$, we may, using the last result of equation (8) and assuming that the golf ball has uniform density, write the speed with which the golf ball rolls on leaving the clubface as $\tfrac{5}{7}V\sin\theta$.

Actually, from many photographs in golf literature, we know that the golf ball compresses a significant amount, decreasing the distance from the surface of the ball against the clubface to the center of the ball. Equation (8) should be modified using a variable radius $R(t)$ to calculate the torque, to read

$$\int F(t)R(t)dt = 0.4mRV_2$$

$$< \int F(t)Rdt = R\int F(t)dt = Rm(V_1 - V_2). \tag{10}$$

Equation (10) tells us that the actual speed of the ball up the clubface at the end of the collision is $\tfrac{5}{7}fV\sin\theta$, where f is the fraction of the speed remaining when the compression of the ball is considered. We do not know the exact value of the fraction f.

The diagram in Fig. A3.4 shows the vectors for the collision between a clubhead and a golf ball when realistic values of e, a, and f are used. Three vectors for the ball velocity up the clubface at the end of the collision are shown: $V\sin\theta$ is the component of the ball's velocity up the clubface for the case of frictionless sliding, $\tfrac{5}{7}V\sin\theta$ is the component of this velocity up the clubface for the case of a uniform rolling

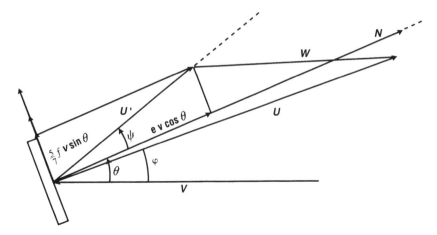

FIGURE A3.4. Drawn to scale in the moving frame for a collision of a ball and a clubhead when there is friction between the ball and the clubface. The coefficient of restitution is taken as 0.7, and the ratio of the mass of the ball to that of the clubhead is 0.19. The vectors N, V, U and the angles θ and φ have been described in Fig. A3.1. The vector U' represents the velocity of the ball after the collision in the moving frame, and it is at an angle ψ to the vector N. Notice that the vector U is in the rest frame and is the sum of U' in the moving frame and W, the velocity of the moving frame relative to the rest frame.

ball that is not compressed, and $\frac{5}{7}fV \sin \theta$ is the component of this velocity up the clubface for a ball that has been compressed against the club face. The latter component of the velocity is the most realistic of the three. In this case, the velocity of the ball in the moving frame as it leaves the clubhead surface after the collision is the sum of its normal component, $eV \cos \theta$, and its component along the clubface, $\frac{5}{7}fV \sin \theta$. This velocity vector is indicated by U' in Fig. A3.4. This velocity vector U' is at an angle ψ from the normal vector N. The tangent of the angle ψ is seen to be

$$\tan \psi = \left(\frac{5f}{7e}\right) \tan \theta. \qquad (11)$$

The velocity U of the ball in the rest frame after the collision is obtained by adding the velocity of the moving frame W to the velocity of the ball in the moving frame. From the figure, which is drawn to scale for $e = 0.7$, $a = 0.19$, and $f = 0.7$, with $\theta = 21$ deg, the loft for a two iron, it is seen that the velocity vector U does lie below the normal vector N.

We observe that for making calculations of the angle φ within the accuracy allowed by this theory, a further equation may be obtained by applying the trigonometric sine law to the triangle with sides U', W, and U. We find that

$$\frac{W}{U} = \frac{\sin(\theta + \psi - \varphi)}{\sin(\theta + \psi)}. \tag{12}$$

Equations (3) and (4) give

$$\frac{W}{U} = \frac{\cos(\theta - \varphi) - ae\cos\theta\cos\varphi}{(1 + e)\cos\theta}. \tag{13}$$

Combining equations (12) and (13), we have

$$\frac{\cos(\theta - \varphi) - ae\cos\theta\cos\varphi}{(1 + e)\cos\theta} = \frac{\sin(\theta + \psi - \varphi)}{\sin(\theta + \psi)}. \tag{14}$$

We have seen that $\tan\psi = (5e/7f)\tan\theta$. We may use equation (14) to solve for $\tan\varphi$ in terms of $\tan\theta$. The algebraic expression is

$$\tan\varphi = \frac{e(1 + a)(1 + 5f/(7e))\tan\theta}{\tan^2\theta(1 - 5f/7) + (1 + e)}. \tag{15}$$

For $\theta = 21\deg$, $a = 0.19$, $e = 0.7$, and $f = 0.7$, equation (15) gives $\varphi = 17.1\deg$.

Since we cannot specify exact values for e and a, we find that we are unable to calculate the trajectory angle φ that a given golfer will achieve with a given club. We are able, however, to consider the individual effects on the angle φ for the various factors describing a given collision between the clubhead and the ball.

Any increase in the angle θ, the angle between the velocity vector V of the clubhead just before impact and the normal N to the clubface, will obviously increase the angle φ. There are three factors determining the angle θ for a given swing. The loft of the club is the dominant factor. The orientation of the clubface will be affected by the position of the shaft of the club at impact. When the hands of the golfer are ahead of the ball at impact, the angle θ will be decreased. The golfer uses this fact to hit a lower or higher shot. The flex of the shaft of the club will have an effect on the angle θ. Usually, the shaft is flexed at impact, so that the normal N to the clubface is raised and thus the angle θ is increased somewhat. A club with a very flexible shaft should send the ball on a higher trajectory than one with a stiffer shaft. The loft of the club is seen to be but one factor in determining the angle φ.

The factor a, the ratio of the mass of the ball to the effective mass of the clubhead, affects the trajectory angle φ through its effect on the velocity vector W, the relative velocity of the moving frame and rest frame after collision. Equation (3) shows that a smaller value of a (a larger clubhead mass) produces a larger value of W. As may be seen from Fig. A3.4, a larger value of W produces a smaller value of φ. A club with a smaller clubhead mass will send the ball on a higher trajectory than the same club swung in the same way with a larger clubhead mass.

The effect of e, the coefficient of restitution, and f, the quantity descriptive of the effect of the compression of the ball, appear as the ratio f/e in the expression for the angle ψ of rebound in the moving frame. This ratio will depend on the nature of the golf ball and on the velocity vector V of the clubhead. A hard ball, one with high compression, will have a larger coefficient of restitution and also a larger value of f than a softer ball. Both the value of e and the value of f will decrease as the clubhead velocity V increases. The fraction 5/7 in the expression for the angle ψ holds only for a sphere of uniform density. For a ball with a greater density toward the center, this fraction will be greater and the trajectory angle φ will also be greater. This tangle of uncertainties makes it impossible to use this bit of theory for quantitative calculations of the trajectory angle φ after the collision. Experience has shown, according to expert golfers, that increases in clubhead velocity and using a ball of higher compression both work toward decreasing the angle φ.

Our model of what happens between the ball and the clubface may be at fault because we have assumed that the ball is rolling up the clubface when the ball and clubface separate at the end of the collision. We know that the ball is rotating because no normal shot is without some lift. It is generally known that balls hit from deep rough have large trajectory angles. This effect is caused by the ball sliding up the clubface with a velocity closer to $V \sin \theta$, the velocity along the clubface when there is no friction.

If the ball is indeed rolling when it leaves the clubface, the angular velocity of the ball is determined as in the case of the bowling ball. Some golfers have found that harder balls do not have the lift of softer balls and thus have less angular velocity. If this interpretation is correct, it follows that with the harder balls the force of sliding friction is not large enough for the ball to be rolling on leaving the clubface.

There seems to be an incentive by all almost all golfers to hit the ball farther. One approach to this goal is to learn how to put more energy into the clubhead, and thus obtain a greater clubhead speed when the clubhead hits the ball. Another is to change the equipment so that less energy is lost in the collision between the clubhead and the ball.

The coefficient of restitution, e, depends on this energy loss. The quantity e for a golf ball is limited by the rules of golf. The velocity of a golf ball shall not be greater than 250 feet per second when hit by a machine devised by the USGA. Balls may be made which are more elastic than those allowed by this rule of golf, any game in which these balls are used is simply not golf.

The overall coefficient of restitution in the collision of the clubhead and the ball may be increased by changing the design of the clubhead. The idea involved in this process is called the "trampoline effect." When the face of the club is made of a thin metallic sheet, it should, according to proponents of the process, cause less energy to be lost in the collision of the clubhead and the ball. The rules of golf would then probably be changed to forbid the use of such clubs in the game of golf.

However it is of interest to look at equations 1, 2, 3, and 4 in this Section, and their use in the energy equation of the collision of the clubhead with the ball. We shall simplify this consideration by having the actual loft of the club to be zero and having the golfer swing the club so that the effective loft is also zero. For such a collision the ball will have no spin and thus the kinetic energy of the spin of the ball will also be zero. With angles θ, and φ, thus both zero, the cosines of these angles will be unity. The equations 1, 2, 3, and 4, become

$$V = W + aU \tag{1'}$$

$$W = U - eV \tag{2'}$$

$$W = V(1 - ae)/(1 + a) \tag{3'}$$

$$U = V(1 + e)/(1 + a) \tag{4'}$$

The kinetic energy $\frac{1}{2}MV^2$ of the clubhead before impact is divided into three parts during the collision. The kinetic energy of the clubhead after the collision is then $\frac{1}{2}MW^2$ and the kinetic energy of the ball is then $\frac{1}{2}mU^2$. We shall represent the energy turned into noise and heat in the collision by the letter E. The theory of conservation of energy then allows us to write

$$\tfrac{1}{2}MV^2 = \tfrac{1}{2}MW^2 + \tfrac{1}{2}mU^2 + E \tag{5'}$$

These equations, after mathematical development, show that the fraction of the clubhead energy before impact left in the clubhead after the collision is

$$(1 - ae)^2/(1 + a)^2 \tag{6'}$$

The fraction of the energy transferred to the ball in the collision is

$$a(1 + e)^2/(1 + a)^2 \tag{7'}$$

The fraction of the clubhead energy before the collision which is lost during the collision is

$$a(1 - e^2)/(1 + a) \tag{8'}$$

When the values of a and e used before in this section are substituted in these three equations we find that the fraction of the original clubhead kinetic energy remaining in the clubhead is .531. The fraction of that original energy in the ball after the collision is .388 and the fraction of that original energy lost in the collision is .081. This analysis of energy transfer during the collision of the clubhead with the ball may be of use for anyone thinking of the design of a clubhead to change the coefficient of restitution of such a collision.

Technical Appendix— Section 4

The Differential Equations

The Two-Rod Model

Any swing of a golf club is a unique dynamical process. While the golf swing in all its complexity and variety is beyond exact analysis, the application of Newton's laws of motion to a simple model of a swing should bring some understanding of what happens in any swing. In this discussion we shall begin with a two-rod model in which the two rods are inflexible. The almost universal present-day use of a straight left arm in swinging the club and the mathematical simplicity of an inflexible shaft justify this approach. Later, we shall look at a model with a flexible shaft. The technical details of a three-rod model, in which the left arm bends at the elbow, are so tedious that they will not be presented here.

When we watch a competent golfer swing a club, we notice that his arms swing about an axis that moves during the swing, and the club swings about an axis near the wrists of the golfer. We shall first choose a mechanical model with the following elements. The model will consist of two rods, rod A representing the arms of the golfer and rod C representing the club, joined at a hinge representing the wrists of the golfer with the upper end of rod A constrained to rotate about an axis that moves horizontally during the downswing. Such a model is shown in Fig. 2.3 and Fig. A4.1. During the downswing the axis at the upper end of rod A will be assumed to move with an acceleration a, first positive toward the target and later negative as it slows and stops.

The golf club is swung in the Earth's gravitational field. The effect of this field is equivalent to the effect of an upward acceleration g of the Earth's surface if it were in a hypothetical gravitational-field-free region. Conversely, the effect of an upward acceleration is equivalent to that of a hypothetical gravitational field in a direction opposite to the acceleration. In our model with the horizontal acceleration a, we may think of the golf club as swung in a field-free region where the axis of

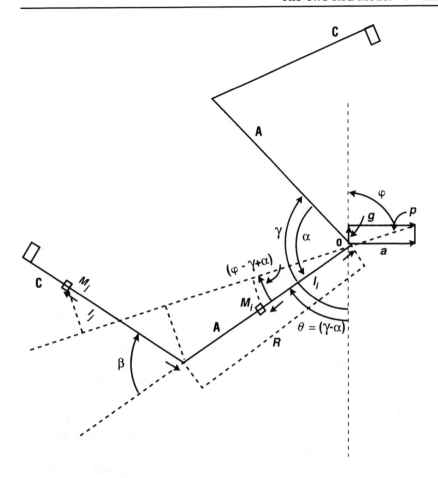

FIGURE A4.1. Diagram of the two-rod model of the golf swing. Rod A represents the golfer's arms, and rod C represents the golf club. At the top of the back-swing, the arm A is at an angle γ with the vertical. After the arm has moved in the downswing through an angle α, the arm is at an angle $\theta = \gamma - \alpha$ from the vertical, and the wrist-cock angle is β. The system swings about a center at O, which moves horizontally in the shift of the golfer. The vectors g and a, with a resultant p, are the acceleration of gravity and the acceleration of the horizontal shift, respectively. R is the length of rod A, I_i is the distance of a mass element m_i of rod A from O, and I_j is the distance of a mass element m_j of rod C from the wrist-cock axis. This diagram is useful in calculating the potential energy of the system in any position in the downswing.

the swing has an acceleration p, which is the vector sum of a horizontal acceleration a and an upward acceleration g, or we may think of it as swung in a hypothetical gravitational field acting in a direction opposite to the acceleration p. We shall adopt the second point of view.

The Use of the Lagrangian Method

Newton's laws of motion can be applied to this model of the golf swing most easily only in an indirect way; even the two-rod model is too complicated to allow the differential equations of motion to be written directly. We shall therefore use the Lagrangian method. In this method we must have expressions for the potential energy and the kinetic energy of the system during the downswing. In arriving at the expressions for these energies we shall calculate them for small mass elements of the system and sum them to obtain the energies of the whole system. We shall write the Lagrangian function that is the difference between the kinetic energy and the potential energy of the system expressed in suitable generalized coordinates. The Lagrangian equations will then be used to arrive at the differential equations of motion in terms of these generalized coordinates and generalized forces.

The generalized coordinates are angles. In Fig. A4.1 the angle γ (gamma) shall represent the backswing angle of rod A. The angle α (alpha) shall represent the angle through which rod A has rotated to any position in the downswing. The angle θ (theta) is equal to $\gamma - \alpha$ and thus represents the angle through which rod A has rotated from the downward direction. The angle β (beta) shall represent the wrist-cock angle measured clockwise from the extension of rod A. These angles are all measured in the plane of the swing.

As shown in Fig. A4.1, we shall think of rods A and C as divided into a large number of mass elements m_i and m_j, respectively. The mass element m_i will be at a distance l_i from the axis O, and mass element m_j will be at a distance l_j from the wrist-cock axis, where rods A and C are joined by the hinge, the flexibility of which is under the control of the golfer.

When the horizontal acceleration a of the point O is considered to be constant for some time interval, we may write the expression for the potential energy of the system during that interval in a hypothetical gravitational field with an acceleration equal and opposite to p, the vector sum of a and g.

The Potential Energy of the System

The potential energy of the system is the sum of the energies of all the mass elements of the two rods. From Fig. A4.1, elementary considerations lead to the expression for the potential energy of systems at any time during the swing measured from the center of rotation at O. It is

$$PE = -(SA + MC \cdot R)(g \cos \theta + a \sin \theta) - S[g \cos(\theta + \beta)$$
$$+ a \sin(\theta + \beta)].$$

Here, remembering that rods A and C represent the arms and club, respectively, SA is the first moment of the arms about the axis at O, MC is the mass of the club, R is the length of the arms, and S is the first moment of the club about the wrist-cock axis.

The Kinetic Energy of the System

The kinetic energy of the system is likewise the sum of the kinetic energies of the mass elements in their motion about the axis at the point O. The kinetic energy of m_i is $\frac{1}{2} m_i (l_i \dot{\alpha})^2$, where $\dot{\alpha}$, using the dot notation, is the first time derivative of α. The kinetic energy of m_j is $\frac{1}{2} m_j V_j^2$, where V_j is the velocity of m_j relative to the system of coordinates in which the axis at O is at rest. As is shown in Fig. A4.2, the velocity V_j is the vector sum of two velocities; the first is $r_j \dot{\alpha}$, the velocity of m_j in the inertial system when β is constant (r_j is the distance of m_j from O); the second is $-l_j \dot{\beta}$, the velocity of m_j relative to the system of coordinates rotating about O with an angular velocity $\dot{\alpha}$. These velocities are positive when $\dot{\alpha}$ is positive and $\dot{\beta}$ is negative.

From the geometry of Fig. A4.2 it follows that

$$R^2 = r_j^2 + l_j^2 - 2 r_j l_j \cos \mu,$$
$$r_j^2 = R^2 + l_j^2 + 2 R l_j \cos \beta,$$
$$V_j^2 = r_j^2 \dot{\alpha}^2 + l_j^2 \dot{\beta}^2 - 2 \dot{\alpha} \dot{\beta} r_j l_j \cos \mu.$$

From these equations it follows that

$$V_j^2 = [R^2 + l_j^2 + 2 R l_j \cos \beta] \dot{\alpha}^2 + l_j^2 \dot{\beta}^2 - 2[l_j^2 + R l_j \cos \beta] \dot{\alpha} \dot{\beta}.$$

The expression for the kinetic energy of the system is

$$KE = \sum_{ijj} \left(\tfrac{1}{2} m_i l_i^2 \dot{\alpha}^2 + \tfrac{1}{2} m_j V_j^2 \right),$$

where the summation extends over all the mass elements m_i and m_j. This expression takes the form

$$KE = \tfrac{1}{2}[J + I + MC \cdot R^2 + 2 RS \cos \beta] \dot{\alpha}^2 + \tfrac{1}{2} I \dot{\beta}^2$$
$$- [I + RS \cos \beta] \dot{\alpha} \dot{\beta}.$$

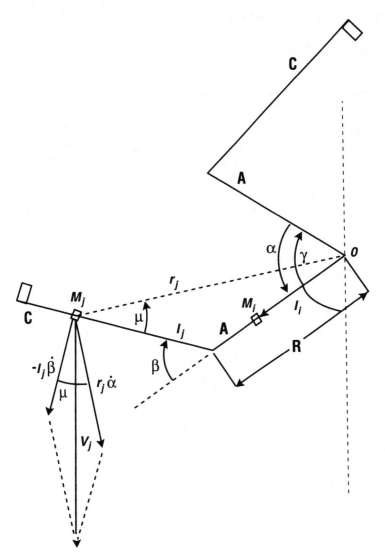

FIGURE A4.2. Similar to Fig. A4.1. This figure is used to calculate the velocity of m_i and m_j during the downswing so that the kinetic energy of the system may be calculated. The vector V_j, the velocity of m_j in the stationary frame, is the sum of two vectors: $-I_j\dot{\beta}$ and $r_j\dot{\alpha}$; $r_j\dot{\alpha}$ is the velocity of m_j in the stationary frame when β is constant, and $-I_j\dot{\beta}$ is its velocity in the rotating frame when β changes. The velocity of m_j is simply $I_j\dot{\alpha}$. The distance r_j is that of m_j from O.

Here $J = \Sigma m_i l_i^2$ and $I = \Sigma m_j l_j^2$ are the second moments of rods A and C about their respective axes.

The Lagrangian Function of the System

With α and β as the generalized coordinates, the Lagrangian function $L = \text{KE} - \text{PE}$ has the form

$$L = \tfrac{1}{2}[J + I + MC \cdot R^2 + 2RS \cos \beta]\dot\alpha^2 + \tfrac{1}{2}I\dot\beta^2$$
$$- [I + RS \cos \beta]\dot\alpha\dot\beta + [SA + MC \cdot R][g \cos \theta + a \sin \theta]$$
$$+ S[g \cos(\theta + \beta) + a \sin(\theta + \beta)].$$

The Lagrangian Differential Equations of Motion

The Lagrangian differential equations are

$$\frac{d}{dt}\left(\frac{\partial L}{\partial \dot\alpha}\right) - \frac{\partial L}{\partial \alpha} = Q_\alpha, \quad \frac{d}{dt}\left(\frac{\partial L}{\partial \dot\beta}\right) - \frac{\partial L}{\partial \beta} = Q_\beta,$$

where Q_α and Q_β are the generalized forces. The generalized forces are forces that do work on the system when the generalized coordinates increase. Since the generalized coordinates are angles, these generalized forces are torques.

The Lagrangian differential equations take the form

$$(J + I + MC \cdot R^2 + 2RS \cos \beta)\ddot\alpha - (I + RS \cos \beta)\ddot\beta + (\dot\beta^2$$
$$- 2\dot\alpha\dot\beta)RS \sin \beta - S[g \sin(\theta + \beta) - a \cos(\theta + \beta)] - (SA$$
$$+ MC \cdot R)(g \sin \theta - a \cos \theta) = Q_\alpha, \tag{1}$$

$$I\ddot\beta - [I + RS \cos \beta]\ddot\alpha + \dot\alpha^2 RS \sin \beta + S[g \sin(\theta + \beta) - a \cos(\theta$$
$$+ \beta)] = Q_\beta. \tag{2}$$

The differential equations of motion will tell us what happens in a particular swing, according to our model, when the torques Q_α and Q_β, the acceleration a, and the boundary conditions are specified. The various parameters of the arms and club are assumed known. The boundary conditions are the various angles α, β, and γ and the angular velocities $\dot\alpha$ and $\dot\beta$ at the beginning of the downswing.

The Qualitative Analysis of the Differential Equations

From the differential equations we can learn some things about the golf swing that will help us when we come to actually solve them for quantitative results.

Consider writing

$$TH = (I + RS \cos \beta)\ddot{\alpha} - RS\dot{\alpha}^2 \sin \beta - S[g \sin(\theta + \beta)$$
$$- a \cos(\theta + \beta)].$$

Then equation (2) becomes

$$I\ddot{\beta} = TH + Q_\beta. \tag{3}$$

Remember that Q_β is any outside torque applied to the club, and it does work when the coordinate β increases. Let us consider Q_β equal to zero temporarily. Equation (3) then tells us that $\ddot{\beta}$ is not equal to zero. We notice that for a while during the downswing $\ddot{\beta}$ will be positive, since $\ddot{\alpha}$ is positive, the second term has only a small negative value, since $\dot{\alpha}^2$ is small, and the third term is positive for small values of a. The term TH is thus positive at the start of the downswing. If $\ddot{\beta}$ is positive, the club will swing toward larger values of the wrist-cock angle inward toward the neck of the golfer. Photographs show that the golfer does not allow this to happen. He keeps his wrist-cock angle constant, or nearly so, well into the downswing. To do this he must apply a torque $Q_\beta = TC$ by his wrists. Equation (3) becomes

$$I\ddot{\beta} = TH + TC,$$

and if $TC = -TH$, we have $\ddot{\beta} = 0$. This torque $TC = -TH$ does no work, since β does not change if $\dot{\beta} = 0$ and $\ddot{\beta} = 0$.

The torque TH decreases during the downswing as the negative term $- RS\dot{\alpha}^2 \sin \beta$ becomes larger. When TH becomes negative, the centrifugal torque in the rotating system, if there is no other torque, acts on the club to produce a negative $\ddot{\beta}$ and hence a decreasing wrist-cock angle β. Thus the club hits the ball by itself if the golfer simply relaxes his wrists when the torque TH becomes zero and keeps them relaxed thereafter.

The golfer may exert a torque other than $TC = -TH$ by his wrists during the downswing. He may exert a torque TE (E for extra) in addition to the torque $- TH$. If such a torque is positive, it will make $\ddot{\beta}$ positive, and since $\dot{\beta}$ is zero at the start of the downswing, the wrist-cock angle will be increasing, and if the torque is negative, it will be decreasing. Using $TC = -TH + TE$ we may study the effect of any wrist action on the swing of the club.

We may consider the downswing in two distinct phases. From the start of the downswing Phase I continues until TH becomes zero. During this phase we shall put $TC = -TH + TE$. If TE is kept equal to zero in this phase, $\ddot{\beta}$ is equal to zero; this means that the original wrist-cock angle is maintained throughout this phase of the downswing. If TE is a positive torque, then $\ddot{\beta}$ is greater than zero; this means that the wrist-cock angle will be increasing during this phase. If TE is negative, it will be decreasing.

Phase II of the downswing starts when the torque TH becomes zero and lasts until the ball is hit. During the second phase we may put $TC = TE$. If TE is equal to zero, the wrists are flexible and thus have no effect on the uncocking process. If TE is positive, the golfer, through his wrist action, is hindering the uncocking process, and if TE is negative, his wrist action is helping the uncocking process. During both phases $Q_\beta = TC$. The torque on the arms during the downswing Q_α is the sum of two torques (TS and TE). The torque TS is the torque by the golfer on his arms, and the torque TE is the reaction on the arms from the wrists exerting a torque TE on the club. Thus $Q_\alpha = TS + TE$; the plus sign comes about because α and β are measured in opposite senses.

By choosing the torques TS and TE, both of which may vary during the downswing, and β and γ at the top of the backswing, and using these values in solving the differential equations, we may calculate all kinds of swings of any club. We may put a and g equal to zero to find the effect of these terms on any particular swing.

Solving the Differential Equations

Before we may apply the differential equations to a swing of a particular club by a particular golfer, we must determine the three dynamical parameters of the club, M, C, S, and I, and estimate the effective length R of the golfer's arms and the effective first moment SA and the second moment J of his arms. These estimates are at best not very accurate, and during a calculation they may be varied to obtain the best fit of the calculated swing with the experimental data of the swing.

A brief description how the differential equations are solved is in order. We begin with given values of α and β at the start of the downswing, $\alpha = 0$ and $\beta = \beta(0)$. The equations are used to calculate the values of α and β a short time later into the downswing. These new values are then used as starting values to calculate α and β a short time further into the downswing. This process is repeated many times until values of α and β are obtained that indicate that the ball is hit by the clubhead. A computer is programmed to make these many repetitive calculations. Besides producing the sequence of values of α and β as they depend on time into the downswing, the computer may be programmed to give the speed of the clubhead, the speed of the golfer's hands, and the speed of any other object such as a piece of reflecting tape, Tape A, on the shaft of the club at a known position near the clubhead.

Attention was focused on this tape because it could be clearly seen on the stroboscopic photograph throughout the swing to determine its location and speed as a function of time, whereas the offset of the clubhead made its center of mass difficult to determine with suitable precision during the downswing. The speed of Tape A determined from the stroboscopic photograph as it depends on time into the downswing is shown in Fig. 2.2. The usefulness of our model and the calculations using the differential equations depends on whether, by choosing suitable parameters and torques, a calculated swing can be obtained that fits the experimental data with suitable precision. Indeed, it was found that the experimental curve of the speed of Tape A could be matched almost exactly with such a calculated curve for Tape A. The calculated swing that most nearly fits the experimental data is called the standard swing. This swing is calculated for an inflexible shaft.

The various parameters used in the calculation were adjusted to produce a calculated swing that very closely fits the experimental data. The three dynamic parameters were measured for the club used in the photographed swing. The acceleration of gravity is known. The values of other parameters are estimations. Exact values of $\beta(0)$ and γ cannot be obtained from the photograph, but their sum can be determined with considerable precision. Both $\beta(0)$ and γ were varied in search of the standard swing, but their sum was kept at the measured value. In the standard swing TE is zero, and TS takes on one value throughout the downswing.

The Flexible Shaft Model

A fairly detailed study was made of the flexible shaft model of the swing of the golf club. It was found that a satisfactory simplification of this model could be made by replacing the flexible shaft by a stiff shaft with all the flexing put in a spring just below the grip. The strength of the spring was chosen so that the club had the same frequency of oscillation as the real shaft had when clamped in a vice at the lower end of the grip.

When this shaft was displaced through an angle D, there was a restoring torque $- kD$. The differential equation of motion of the shaft clamped in a vice, neglecting damping, is

$$I\frac{d^2D}{dt^2} = -kD.$$

The solution of this equation is

$$D = \hat{D}\sin\omega t$$

with $k = I\omega^2$. The potential energy of this system is $0.5I\omega^2D^2$. Here ω is 2π times the frequency of the oscillation, and I is the moment of inertia of the club about the point 5 in from the grip end of the club.

From considerations of Fig. A4.3 the total potential energy of the system is

$$PE = -(SA + MC \cdot R)(g\cos\theta + a\sin\theta) - S[g\cos(\theta + \beta + D)$$
$$+ a\sin(\theta + \beta + D)] + \tfrac{1}{2}I\omega^2D^2.$$

The kinetic energy is obtained in the same way as in the previous case of the two-rod stiff shaft model except that the angle β is replaced by the angle $\beta + D$. We obtain

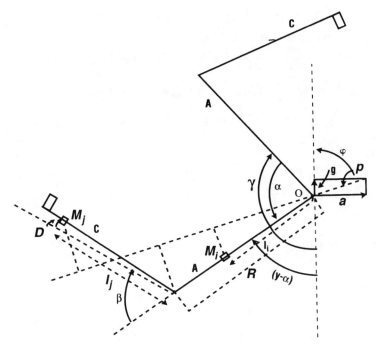

FIGURE A4.3. The same as Fig. A4.1 in most particulars. It shows the angle D, indicating the position of the clubhead from where it would be if the shaft were not flexible. This figure is used to calculate the potential energy of the system when the club is flexible.

$$KE = \tfrac{1}{2}[J + I + MC \cdot R^2 + RS \cos(\beta + D)]\dot{\alpha}^2 + \tfrac{1}{2}[I(\dot{\beta} + \dot{D})^2]$$
$$- \dot{\alpha}(\dot{\beta} + \dot{D})[I + RS \cos(\beta + D)].$$

The Lagrangian is, as before, $L = KE - PE$.
The differential equations are

$$\frac{d}{dt}\left(\frac{\partial L}{\partial \dot{\alpha}}\right) - \frac{\partial L}{\partial \alpha} = Q_\alpha,$$

$$\frac{d}{dt}\left(\frac{\partial L}{\partial \dot{\beta}}\right) - \frac{\partial L}{\partial \beta} = Q_\beta,$$

$$\frac{d}{dt}\left(\frac{\partial L}{\partial \dot{D}}\right) - \frac{\partial L}{\partial D} = Q_D.$$

We have then the following differential equations:

$$[J + I + MC \cdot R^2 + 2RS \cos(\beta + D)]\ddot{\alpha} + (\ddot{\beta} + \ddot{D})[I + RS \cos(\beta$$
$$+ D)] - \dot{\alpha}(\dot{\beta} + \dot{D})[2RS \sin(\beta + D)] + (\dot{\beta} + \dot{D})^2[RS \sin(\beta$$
$$+ D] - (SA + MC \cdot R)(g \sin \theta - a \cos \theta) - S[g \sin(\theta + \beta$$
$$+ D) - a \cos(\theta + \beta + D)] = Q_\alpha,$$

$$I(\ddot{\beta} + \ddot{D}) - [I + RS \cos(\beta + D)]\ddot{\alpha} + RS \sin(\beta + D) \cdot \dot{\alpha}^2$$
$$+ S[g \sin(\theta + \beta + D) - a \cos(\theta + \beta + D)] = Q_\beta,$$

$$I(\ddot{\beta} + \ddot{D}) - [I + RS \cos(\beta + D)]\ddot{\alpha} + RS \sin(\beta + D)\dot{\alpha}^2 + S[g \sin(\theta$$
$$+ \beta + D) - a \cos(\theta + \beta + D)] + I\omega^2 D = Q_D.$$

For this case TH becomes

$$TH = [I + RS \cos(\beta + D)]\ddot{\alpha} - RS \sin(\beta + D) \cdot \dot{\alpha}^2 - S[g \sin(\theta + \beta$$
$$+ D) - a \cos(\theta + \beta + D)].$$

The differential equations for β and D become

$$I\ddot{\beta} = TH - I\ddot{D} + Q_\beta,$$
$$I\ddot{D} = TH - I\ddot{\beta} - I\omega^2 D + Q_D.$$

As in the case of the stiff shaft model, $Q_a = TS$ and $Q_\beta = TC$. But in this case we have Q_D as a damping torque factor, which is expressed as $-DAM*D$.

The process of solving these differential equations follows in the same way as in the case of the stiff shaft model.

TECHNICAL APPENDIX— SECTION 5

Torques and Energies

Finding Energies of Parts of the System

We may increase our understanding of what is happening in the swing of a golf club by examining the details of how energies enter the system and how their distribution in the swing develops. The consideration of the individual torques acting on the arms and club provide further insight into what is happening in this process.

The kinetic energy calculated by the expression used in setting up the Lagrangian function,

$$\tfrac{1}{2}[J + I + MC \cdot R^2 + 2RS \cos \beta]\dot\alpha^2 + \tfrac{1}{2}I\dot\beta^2 - [I + RS \cos \beta]\dot\alpha\dot\beta,$$

may be separated into two parts:

$$\tfrac{1}{2}J\dot\alpha^2$$

and

$$\tfrac{1}{2}[I + MC \cdot R^2 + 2RS \cos \beta]\dot\alpha^2 + \tfrac{1}{2}I\dot\beta^2 - [I + RS \cos \beta]\dot\alpha\dot\beta.$$

The first term is the kinetic energy of the arms in the swing motion. It depends only on the mechanical properties of the arms; J is the second moment of the arms about the axis of rotation. The second term depends on the mechanical properties of the club; I is its second moment about the wrist-cock axis, MC is its mass, S is its first moment about the same axis. Finally, R is the length of the golfer's arms and determines the path on which the wrist-cock axis moves. The kinetic energy associated with the shift motion does not appear in these expressions. These energies for the standard swing are plotted against the angle alpha in Fig. 5.1.

Since the product of torque and angle is work done by the torque turning a system through an angle, the equation Energy=Torque ×Angle, for a constant torque, represents a straight line in the figure; the slope of the line represents the torque involved. Since in the standard swing the torque the golfer exerts on the system is constant, the energy he puts into the system is represented by a straight line. The line D in Fig. 5.1 is plotted with a slope representing the constant torque TS.

After we have solved the differential equations, equations (1) and (2), and have obtained the standard swing fitting our golfer's swing, we may calculate the values of these two expressions for the kinetic energy of the arms and the club. These results are also plotted in Fig. 5.1.

Curve B shows the kinetic energy of the arms as it depends on the angle alpha. At the beginning of the swing, with small values of alpha, curve B follows line D fairly closely. Later during the swing, curve B reaches a maximum and then decreases until the ball is hit.

Curve C is a plot of the kinetic energy of the club as it depends on alpha. At the beginning of the downswing this curve rises somewhat more slowly than either Curve D or Curve B and gradually sweeps to high kinetic energy values.

Curve A is a plot of the sum of the kinetic energies of the arms and the club. Curve A is not a straight line. Curve A lies above line D, and this indicates that the total kinetic energy is greater than the energy supplied by the golfer to the system through the torque TS.

These curves show that within our model the energy lost by the arms as they slow down in the latter part of the swing appears as energy in the club and that the sum of these energies is greater than that supplied by the torque TS on the system by the energy gained from the gravitational field and from the work done by the golfer in his shift of the center of rotation.

Finding Torques in Arms

In the further study of the anatomy of the swing, we may investigate the individual torques that act during the downswing in the golf stroke. On examining the two differential equations, equations (1) and (2), we find five expressions for torques in each of them. For the torques on the arms, they are

$$+ (I + RS \cos \beta)\ddot{\beta},$$

$$- (RS \sin \beta)(\dot{\beta}^2 - 2\dot{\alpha}\dot{\beta}),$$

$$+ g[S \sin(\theta + \beta) + (SA + MC \cdot R)\sin \theta],$$

$$- a[S \cos(\theta + \beta) + (SA + MC \cdot R \cos \theta],$$

$$+ Q_\alpha.$$

A similar list will be given below for the torques on the club.

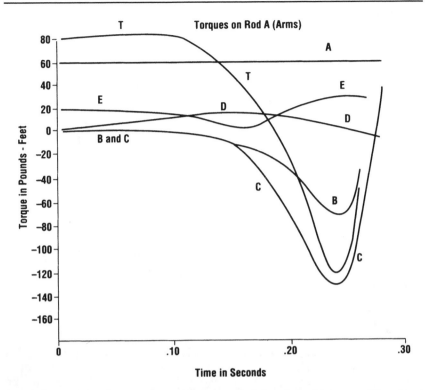

FIGURE A5.1. The five torques acting on rod A (arm) as they vary throughout the downswing. Curve A shows the constant torque TS of the golfer on the system. Curve B shows the torque $(I + RS \cos \beta)\ddot{\beta}$, a torque that depends mainly on the acceleration of the angle β. Curve C shows the torque $- RS \sin \beta(\dot{\beta}^2 - 2\dot{\alpha}\dot{\beta})$, a torque that depends mainly on the square of the velocity of the angle β. Curves D and E show the torques resulting from action of gravity and the golfer's shift, respectively. The torque T shows how the sum of these five torques on rod A varies during the downswing. Notice that this torque is negative after 0.19 s into the downswing and becomes very large before the ball is hit.

The computer was programmed to calculate the terms in both lists. The torques on the arms as they depend on time are plotted in Fig. A5.1. The horizontal line A represents the constant torque $Q_\alpha = TS$ of the golfer on the system. The torque $(I + RS \cos \beta)\ddot{\beta}$ is plotted as curve B. Since $\ddot{\beta}$ is zero in the first part of the swing, the wrist cock is kept constant; this curve starts off along the horizontal time axis and only later plunges to negative values when the wrist is uncocking.

The second torque on the list, plotted in curve C, also starts off along the time axis because when the wrist is not allowed to uncock, the angular velocity $\dot{\beta}$ is zero. Later in the swing the curve plunges to negative values because $\dot{\alpha}$ is positive and $\dot{\beta}$ is negative.

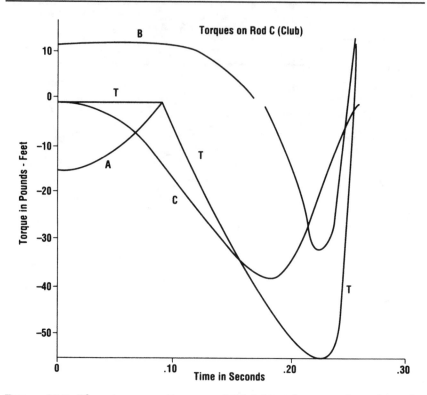

FIGURE A5.2. Three torques acting on rod C (club) as they vary throughout the downswing. Torque A is the torque that the wrists must exert on the club during the first tenth of a second into the downswing to keep the wrist-cock angle constant. Torque B $(I + RS \cos \beta)\ddot{\alpha}$ is the torque on the club produced by the acceleration of the golfer's arms. Torque C, $- (RS \sin \beta)\dot{\alpha}^2$, is the torque on the club depending mainly on the square of the angular velocity $\dot{\alpha}$. This is the centrifugal torque on the club in the rotating system. The torque T is the sum of torques A,B,C, and the two insignificant torques of the club from the action of gravity and the golfer's shift. The last two torques are not shown in the figure.

Curves D and E are, respectively, plots of the torques resulting from the pull of gravity and the shift of the golfer during the swing.

The sum of these five torques acting on the golfer's arms to give them the angular acceleration alpha is plotted in curve T. This curve is nearly constant for about one-tenth of a second into the downswing and then decreases through zero at about 0.19 s. The angular velocity of the arms increases up to this time but begins to decrease as this curve plunges into negative values until the ball is hit. Since the golfer's hands have the same angular velocity as the golfer's arms, this means that the golfer's hands slow up for the rest of the swing.

Finding Torques on the Club

The following is the list of the five torques acting on the club to produce an acceleration of the wrist-cock angle beta:

$$+ [I + RS \cos \beta]\ddot{\alpha},$$

$$- [RS \sin \beta]\dot{\alpha}^2,$$

$$- g[S \sin(\theta + \beta)],$$

$$- a[S \cos(\theta + \beta)],$$

$$+ Q_\beta.$$

The torques on the club, except the two small torques produced by the force of gravity and the shift of the golfer, are plotted in Fig. A5.2. The torque $[I + RS \cos \beta]\ddot{\alpha}$ on the club produced by the acceleration of the golfer's arms is plotted in curve B. This torque starts off positive and holds about constant for the first one-tenth of a second into the downswing. Thereafter, this torque decreases and becomes negative. The torque on the club produced by the centrifugal force in the rotating system depends on the square of the angular velocity of the golfer's arms. This curve is shown as curve C. It is a negative torque, which early in the downswing would produce, if acting alone, an uncocking of the wrists. However, at the start of the downswing the sum of the torques plotted in curves B and C is positive and would produce a further cocking of the wrists if it were not matched by a torque by the wrists. This torque, $Q_\beta = TC = -TH$, is plotted in curve A. The torque plotted in curve A contains the negative of the two small torques involving g and a. The sum of the torques plotted in curves A, B, C and the two small torques is equal to zero until just before one-tenth of a second into the downswing. After this time the torque TC remains zero in the standard swing.

The sum of all the torques on the club is plotted in curve T. After one-tenth of a second into the downswing this sum plunges to the astonishing negative value of about -53 lbs ft. It is this torque, represented in curve T, that produces the rapid uncocking of the wrist in a well-developed downswing. The detailed analysis of the terms in the differential equation shows the complex nature of the swing of a golf club.

TECHNICAL APPENDIX— SECTION 6

Treatment of Data

The stroboscopic photograph of a golf swing, a 2 by 2 in slide, was projected on a large sheet of paper on a wall. The images formed by the light reflected by Tape A on the club shaft showed clearly on the paper. Their positions were marked with a fine felt pen. The distances between successive marks were carefully measured. A metric scale was used to ensure precise estimations of fractional units. A yard stick, which lay at the feet of the golfer swinging the club, had two tapes on it at a known separation. The images of these tapes serve in the calibration of distance in the plane of the swing.

The distances between the successive marks were plotted against the running number of the interval starting at the interval just before the ball was hit. The graph so obtained was extrapolated to zero club speed to establish the time $t = 0$ for the swing. These successive distances are proportional to the speed of Tape A, since the light flashes occurred at constant frequency.

The time interval between light flashes depended on the calibration of the Strobotac, which was uncertain. The duration of the downswing, from time $t = 0$ to the end of the last interval before the ball was hit, was 47 times the time between flashes. This number is based on the assumption that the acceleration of the clubhead immediately after the start of the downswing is constant and therefore the speed of Tape A increased linearly with time.

Instead of trusting the calibration of the Strobotac, the duration of the swing.was assumed to be 0.25 s. This duration of the swing is assumed to be the same as those of other professional golfers with the same type of swing, which were determined by other means [6]. The time between flashes is thus 0.005319 s. The distance Tape A moved in this time divided by this time gives the average speed for the interval. This speed is then taken to be the speed at the middle of the interval. These speeds, for four independent sets of measurements, are shown in Table I. The time interval between the speeds given in this table is 0.005319 s.

These speeds were plotted on a large graph and a smooth curve was drawn through the points. This curve was extrapolated to zero speed to determine the time of the start of the downswing. The averages of the speeds shown in Table 1 are plotted as dots in Fig. 2.2A.

The standard swing devised to match the original data is shown in Table II. Column A is the time in seconds, column B shows the smoothed observed data, and column C shows the calculated speeds of Tape A.

This discussion shows why the reader should not think of numbers in the text as describing exactly what is happening during the swing of the club.

TECHNICAL APPENDIX— SECTION 7

Stroboscopic Photograph of the Golf Stroke

Detailed information of what happens in a particular golf stroke may be obtained by photographing the stroke with the flashing light of a stroboscope. The usual photographs seen in the golf literature are made with a light so intense that each flash would give a suitably exposed photograph of the golfer, and a photograph made with many flashes shows a very blurry effect of many superposed pictures. With the use of reflecting tape attached to specific places on the club and on the golfer and with the use of a less intense light source, photographs may be made in darkness with a flashing light to show black spots on the negative indicating the position of the bits of tape at each flash of the light. The general illumination from the numerous flashes during the exposure for one stroke is not enough to show the golfer on the photograph.

The tape used was Scotchlight Tape 7610 supplied by the Minnesota Mining and Manufacturing Company (3M). This particular tape "reflects approximately one thousand times as much light as a white painted surface" in a narrow beam back along the direction of the incident light. A 35 mm camera with a 50 mm $f1.4$ lens gave suitably exposed negatives on Kodak Tri-X film, ASA 400, at a distance of 25 ft between the camera and the golfer when a General Radio Strobotac, Type 1531-AB, was used at a flash rate of 120 per second. This rate gives roughly 150 flashes from the time the backswing is begun to the time the ball is hit.

A minimum number of bits of tape should be used on any one photograph. Otherwise, there may be some difficulty in interpreting the resulting negative. At least three pieces of tape should be placed around the shaft of the club, one piece should be placed on the left hand of the golfer in a position that can be seen from the camera throughout the swing, and one piece may be placed on the head of the golfer. Tape should be placed on the head of the club so that it may be seen throughout the swing. Tapes may be placed at other points to study special effects, such as bending the left elbow, but no attempt

168

should be made to try to learn everything about a swing in one photograph. Interesting details of the swing may be obtained by photographing the plane of the swing edge with the tape attached to the head of the club. The scale of the photograph may be established by laying a suitably taped yardstick on the ground just in front of the golf ball on the tee.

The 35 mm negatives may be mounted in slide mounts and projected as negatives with a regular slide projector. Measurements may be made directly from the image projected on a smooth wall, or the image may be projected on a large sheet of paper and details of the image marked on the paper with a fine felt pen. Measurements may be made later from this permanent record.

TECHNICAL APPENDIX—
SECTION 8

Tables

TABLE I. Four independent determinations of the average speed in successive time intervals between flashes of the stroboscope in the photograph of the golf swing. These data are plotted as dots in Fig. 2.2. The speeds are expressed in feet per second. They are determined as explained in the Technical Appendix Section 6.

8.19	8.19	8.19	8.00	42.10	42.10	43.44	42.10
9.34	9.14	9.34	9.38	45.39	44.58	45.53	45.15
9.72	9.91	9.53	9.91	48.01	48.20	48.77	48.96
11.62	11.43	11.43	11.43	53.15	52.77	52.01	53.53
12.57	12.57	13.15	13.34	*	*	*	*
*	*	*	*	*	*	*	*
*	*	*	*	66.49	66.87	67.06	66.69
16.57	15.24	16.38	14.67	72.76	72.39	72.01	73.35
17.91	17.91	17.91	17.72	78.30	78.11	79.63	78.11
18.29	20.96	19.24	18.48	84.21	84.78	83.63	84.02
19.81	21.34	18.86	20.00	90.68	90.49	90.87	91.06
21.15	21.72	21.34	21.53	97.16	96.40	97.35	96.97
22.57	22.48	22.57	22.57	103.07	104.40	103.07	103.07
24.00	23.43	23.05	23.24	109.93	109.35	109.35	109.16
24.39	25.15	25.53	24.77	116.02	116.40	116.40	116.40
26.50	26.86	26.29	26.67	122.69	121.93	121.36	121.93
28.58	28.58	28.77	28.39	127.83	127.83	127.26	126.69
29.91	28.58	30.10	30.10	133.36	134.12	133.74	135.26
31.63	30.10	31.24	31.24	139.46	138.31	138.50	138.50
33.53	33.91	33.53	34.86	142.88	143.07	143.27	143.26
36.72	36.96	36.58	36.00	146.12	144.98	144.98	145.17
39.25	38.48	38.10	38.48				

TABLE II. A comparison between the experimental data and the calculations of the swing from the differential equations for the speed of Tape A. Column A shows the time of the swing in seconds. Column B shows the speed of Tape A obtained at the ends of 0.010 s intervals from the smoothed data presented in Table I. Column C shows these same speeds of Tape A at 0.005 s intervals as calculated for the standard swing.

A	B	C	A	B	C	A	B	C
0.005		1.4	0.105		27.8	0.205		104.8
0.010	2.7	2.8	0.110	29.0	29.3	0.210	110.0	110.5
0.015		4.2	0.115		30.9	0.215		116.4
0.020	5.3	5.6	0.120	31.9	32.7	0.220	121.0	122.1
0.025		7.0	0.125		34.6	0.225		127.5
0.030	8.0	8.3	0.130	36.3	36.8	0.230	131.0	132.4
0.035		9.7	0.135		39.2	0.235		136.8
0.040	10.6	11.1	0.140	42.1	41.8	0.240	140.6	140.6
0.045		12.4	0.145		44.7	0.245		143.7
0.050	13.3	13.7	0.150	48.0	47.9	0.250	146.0	146.1
0.055		15.1	0.155		51.5	0.255		147.8
0.060	16.0	16.4	0.160	55.5	55.3	0.260	146.5	149.1
0.065		17.7	0.165		59.6			
0.070	18.6	18.9	0.170	64.8	64.2			
0.075		20.2	0.175		69.4			
0.080	21.3	21.4	0.180	75.0	74.7			
0.085		22.7	0.185		80.3			
0.090	23.6	23.9	0.190	85.5	86.2			
0.095		25.1	0.195		92.2			
0.100	26.6	26.4	0.200	97.5	98.3			

TECHNICAL APPENDIX— SECTION 9

Newtonian Dynamics

Linear and Rotational Motion

Before we begin to examine the application of physics to golf, we should realize that each of us has an intuitive awareness of certain aspects of the nature of motion. We share this awareness with animals and plants. A colt learning to use its legs seems to know that he must lean into the curve when his antics bring him to move in a sharp turn, and a hawk circling in the summer sky knows instinctively to bank in the turns. An oak tree is certainly influenced by Earth's gravitational field. The physics in each case is very similar. We should be able to draw on our awareness in developing some understanding of the laws of motion and in presenting the vocabulary in general use for discussing the various aspects of motion. Some part of the discussion will be so simple as to appear trivial, while other parts may require some careful consideration to appear at all reasonable. Remember that we did not understand this subject until relatively recently. Many examples of the application of the particular ideas under discussion will be considered so that those without previous scientific study may understand how physics applies to golf.

Speed, Linear Velocity, and Angular Velocity

One important characteristic of motion is how fast it is taking place. There are two important types of motion, linear motion and rotational motion. In linear motion, a body moves along a line, either a straight line or a curved line. The flight of a golf ball is an example of motion along a curved line. In discussing linear motion we use the word velocity to express two aspects of such motion, its speed and its direction. A well-driven ball off the tee travels at a high speed, while a ball during a short putt moves at a low speed. The skill in putting depends in part on learning to give the ball the right speed to make it roll the correct distance.

The velocity of a body can be changed by having its speed changed while the direction of its motion is unchanged. A falling body moves faster and faster as it falls. The velocity of a body can also be changed by having its direction change while its speed remains constant. The tip of a blade on a rotating electric fan has this type of motion. Its speed is constant, but the direction in which it is moving is changing all the time. The velocity of a body can be changing by having its speed and its direction changing at the same time. You see this type of motion whenever you watch the flight of ball in any golf shot; during the flight the speed of the ball is gradually changing, and all the while the direction of its motion is also changing.

In rotational motion a body rotates about some axis. A merry-go-round rotates about a vertical line at its center. We use the term "angular velocity" to indicate how rapidly a body is rotating. The shaft of an electric motor may turn with a large angular velocity, perhaps 1800 turns per minute, whereas Earth in its rotation turns with a small angular velocity of one turn in one day.

A body may have linear velocity and rotational motion at the same time. A stick thrown end over end has linear motion because it is moving from one place to another and has rotary motion because its orientation in space is changing. A well hit golf ball will have a high speed and be spinning at the same time.

Linear and Rotational Acceleration

Under certain circumstances, the velocity of a body may be changing. When this happens we say that it has an acceleration. Acceleration is a measure of how rapidly the velocity of a body changes with time. The dropped ball, the tip of a fan blade, and the ball in flight have motions with acceleration, since in each motion the velocity is changing.

Technically, velocity is defined as the time rate of change of the position of a body, and acceleration is defined as the time rate of change of the velocity of a body. Angular velocity is defined as the time rate of change of rotation of a body about an axis. Angular acceleration is defined as the time rate of change of angular velocity of a body.

In the early days of golf, the nature of motion was not understood. When the motion of the planets in the heavens was caused by the pushing of angels, what could have been the interpretation of the motion of a badly sliced golf ball? Golf had been played for some two hundred years before the genius of Sir Isaac Newton removed the angels and replaced them by his law of universal gravitational attraction and his three laws of motion. This new way of thinking about motion brought the opportunity for a theoretical examination of the swing of a golf club, but perhaps the lack of high-speed computers prevented the application of dynamics to this problem, a delay of some three hundred years. Eminently qualified physicists of the past have been interested in golf.

Newton's Laws of Motion

Newton's law of universal gravitation will not concern us here. We shall use his three laws of motion to sharpen our understanding of what is involved in the swing of a golf club, the collision between club face and the ball, and the flight of the ball.

These three laws of motion are usually stated as follows:

1. A body continues at rest or in uniform motion in a straight line unless it is acted upon by a force.
2. The rate of change of the momentum of a body is proportional to the force acting on it and is in the direction of the force.
3. For every action there is an equal and opposite reaction.

These may appear to be formidable statements, but they really describe in an abstract way well-known aspects of our everyday experience. While these laws may be used to make exact quantitative statements, such as those basic to this research, here we shall discuss them only qualitatively.

The Nature of Force

We shall start by considering the third law first because this law is concerned with the nature of "force." It is sometimes stated; For every force there is an equal and opposite force. Here "action" is translated as "force." Force as used here is an intuitive concept. When you push on a body you exert a force on it. When you pick up a book you exert a force on it. In each case there is an opposite force on your hand. The fundamental idea of the third law is that we always find forces to be two ended. Some examples will help to clarify this idea. Consider a cup of tea on a table. The cup of tea pushes down on the table, and the table pushes up on the cup. These two pushes are the two components of the force pair between the cup and the table, and they are equal in magnitude and opposite in direction. It happens that the cup of tea and the table are at rest. But this is true for all forces. When a fast-moving clubhead hits a golf ball, the force of the clubhead on the ball and the force of the ball back on the clubhead are two ends of a force pair and are equal. One of this force pair makes the ball speed up, and the other makes the clubhead slow down.

Many people find Newton's third law hard to understand. To them it seems to apply to the forces between the cup and the table but not to the forces between the clubhead and the ball. They do not see how a horse can move a wagon if the wagon always pulls back on the horse with the same force the horse pulls on the wagon. But when they recognize that the two ends of the pull are on different objects, one component of the force pair acts on the wagon and the other acts on the horse, they realize that it is one end of the force pair that moves the wagon.

Momentum

Newton's first law defines a property of a body called inertia, which describes what happens to a body when no force act on it; the inertia of a body is said to be measured by its mass. When acted upon by a constant unbalanced force, the body will experience an acceleration proportional to the mass of the body. The mass of a body is proportional to its weight. Momentum is defined as the mass of a body multiplied by its velocity. Like velocity, momentum has direction as well as magnitude. From the definition of momentum, for constant mass the rate of change of momentum is the product of the mass and the rate of change of its velocity, or the product of the mass and its acceleration. Newton's second law tells us that an unbalanced force on a body is associated with its acceleration. For our purposes the second law becomes, the mass of a body multiplied by its acceleration is proportional to the force acting on it, and the acceleration is in the direction of the force.

The force in this statement must be taken to be the total force acting on the body. Actually, in many cases there will be several forces acting on the body. Consider the forces acting on a person on water skis. The tow-rope exerts a force on him. Earth pulls on him; this is his weight. The water skis exert a force on him, and there may be a force of the wind against him. These four forces may add up to zero. In that case there is no unbalanced force, and according to the second law, he will have no acceleration. His velocity will be constant. But if these forces do not add to zero and there is a net force acting on him, the second law tells us that there will be an acceleration. This means that either his speed will be changing or the direction of his motion will be changing, or both may happen. The next time you see a water skier following a boat in a long sweeping turn, think "acceleration."

Since the relation of force to motion cannot be understood without recourse to Newton's second law, we should look at various examples of motion to see how they are interpreted. In looking at these examples we shall attempt only to obtain a qualitative feeling for the use of the second law in furthering our understanding of force and motion.

Consider again the cup on the table. There is no motion. We say that the cup is at rest. The acceleration is zero. This means that according the second law no net force acts on the cup. The forces on the cup are the push upward by the table and the pull of gravity, its weight, downward on the cup. We conclude that these forces add up to a zero force. These two forces are not a third-law pair because they are acting on the same body. The forces of a third-law pair do not act on the same body. Any body that has no acceleration may be looked at in the same manner as we have looked at the cup; the net force on it is zero.

Let us next consider the case of a body that is accelerating. A freely falling body is a good example. When air resistance may be neglected, the acceleration is practically constant. According to the second law, this means that a freely falling body has a constant force acting on it; this force is its weight. On Earth's surface, this acceleration is about 32 feet per second each second. On the Moon it is about 5.5 feet per second each second. A freely falling golf ball moves very nearly with this acceleration, but a feather or a sheet of paper will not fall with a constant acceleration because there is a variable air resistance.

If a person tosses a golf ball to land a few feet from him, this motion is also one with a constant acceleration because only the force of gravity acts on the ball after it has been thrown. It may be shown that the ball moves on a particular curve called a parabola. However, a well-hit golf ball does not move on such a curve because the complicated variable air resistance acts on the ball, and so the acceleration is not constant.

Collisions and the Conservation of Momentum

In a collision between two bodies various things can happen. Noise is usually produced, bodies may have their shapes changed, and their velocities will be changed both in speed and direction. But in all collisions where no outside forces act, the sum of the momentum of the colliding bodies after the collision is the same, both in direction and magnitude, as the sum of the momentum of the bodies before the collision. This principle of the conservation of momentum in collisions can be shown to follow from Newton's laws of motion.

Perhaps the simplest collision to think about is the firing of a gun. If the gun is at rest before it is fired, the momentum of the gun and the bullet are both zero. After the gun is fired and the bullet is on its way, the total momentum of the two is still zero. The momentum of the bullet in one direction is exactly opposite to the momentum of the gun in the opposite direction. Since the mass of the bullet is much smaller than the mass of the gun, the velocity of the bullet is much larger than that of the gun. The stopping of the recoiling gun is what produces the kick of the gun against your shoulder.

Another collision of more interest to the golfer is that between a rapidly moving clubhead and the golf ball at rest. Before the collision, the clubhead may be moving horizontally close to one hundred miles per hour. After the collision, for a club without loft, the ball is moving off at a high velocity, and the clubhead continues in the follow-through at a somewhat reduced velocity. For such a club these velocities will also be horizontal. But these velocities are such that the momentum of the clubhead before the collision is equal to the sum of the momentum of the ball and the momentum of the clubhead after the collision. There is nothing the golfer can do during the collision, which lasts less than one thousandth of a second, to influence these after-collision velocities.

Newton's first law is a special case of the second law. It tells what happens when no net force acts on a body. Apparently, Newton found it necessary to make this statement because he felt that most of his readers still held the ancient belief that a body needed a net force on it in order that it continue to move with constant velocity. We have seen that the property of a body described in the first law, that of continuing to move in a straight line at constant speed when no net force acts on it, is called inertia.

Rotational Motion

Since we are interested in understanding the swing of a golf club and since the golfer's hands and the head of the club move on curves resembling circles, we should consider applying Newton's law to true circular motion. When a body moves in a circle, its velocity is changing continually even though the speed may be constant. On analyzing this motion it is found that there is an acceleration toward the center of the circle. According to the second law there must be a force in the direction of the acceleration. When we look for this force, we indeed find such a force.

If a string is tied to a spool of thread and the spool is whirled in a circle at the end of the string, the string will be found to be under tension. The faster the spool is whirled the greater will be the tension. If you do this little experiment, be careful that the string does not break under this tension. Since the acceleration is toward the center of the circle, the force acting on the spool by the string is toward the center of the circle. This is called a centripetal force. If the string should break, the spool would not continue to move in a circle but would move along a line tangent to the circle on which it had been moving. This is an example of the effect of inertia.

The importance of rotational motion in the golf swing leads us to consider it in some detail. Let us go in our imagination for ride on a merry-go-round. When we do this, we find ourselves moving in a circle with a centripetal acceleration. There must be a centripetal force acting on each of us. We find that we must hang on to something, such as an iron pipe, to supply this force. This is all according to Newton's second law.

Centrifugal Force

But when we ride on a merry-go-round we feel a force, we think, tending to throw us outward from the center. We look outward to see whether there is something out there pulling on us, and we find nothing of the kind. For all the forces we have considered so far there has been something to produce them. We have seen where the force of our weight, the force in lifting a tea cup, and the force to pull a wagon come from. This force arises because we are riding on an accelerating body, the merry-go-round. Such forces, for which we find no origin in objects such as ropes or rods and that do not produce accelerations in the direction of these forces are called by some "pseudo-forces," and this particular one is called a "centrifugal" force.

You may observe this centrifugal force if you place a golf ball on the surface just inside the windshield of your automobile and watch it roll from side to side as you drive around a curve. The ball always rolls to the outside of the curve and rolls more quickly the tighter the turn. Actually, the ball does not accelerate, it appears to accelerate since there is no centripetal force to make it turn in the same path as the car. Its motion is the result of a lack of a centripetal force rather than the result of an outwardly directed force being applied to the ball. I once watched a small boy come running down the aisle in an accelerating subway car. When the car was no longer accelerating and he could walk back to where his mother was sitting, he told her, "I didn't do it, the car made me do it." We would say that the force he felt came because he was riding in an accelerating car.

According to Newton's second law, the centripetal force on a body moving in a circle is proportional to the mass (weight) of a body multiplied by its centripetal acceleration. The centripetal acceleration increases with the radius of the circle on which it moves and with the square of the angular velocity of the motion. The centripetal force and the centrifugal force are not a third-law pair. The force of the string on the spool and the force of spool on the string are a third-law pair.

Rotational Laws of Motion

So far, we have looked at the relation between forces and the linear accelerations that accompany these forces. But in the case of the golf club, the forces applied to the club also produce rotational motion. To understand the relation between these forces and the rotations, we need to look at the laws of rotational motion, which are analogous to Newton's three laws for linear motion. The concepts we use to understand rotational motion are not quite as simple as those we use for linear motion.

Torque

When we attempt to remove the lid on a pickle jar, one hand twists the lid in a counterclockwise sense, and the other hand twists the jar in a clockwise sense, as seen from above. "Torque" is the word used to describe twist in a quantitative manner. If the lid twists off easily, the needed torque is small. If the lid refuses to budge, some means must be found to produce a larger torque.

Torque in rotational motion is analogous to force in linear motion. Torques come in third-law pairs just as forces do. The torque your hand exerts on a jar lid and the torque the jar lid exerts on your hand are a third-law pair. The torque the jar lid exerts on the jar and the torque the jar exerts on the jar lid are another third-law pair.

A torque is determined by two factors. Suppose you are changing a tire and you find that the lug nut is stuck. You proceed by simply increasing the force you apply to the wrench handle. This will increase the torque applied to the nut. Or you may also use a wrench with a longer handle and apply the some force as before. This will also increase the torque on the nut. The torque in this particular case thus depends on the force applied and the length of the wrench handle. The size of any torque is found by multiplying the size of the force by the length of the lever arm, the lever arm being the shortest distance from the line along which the force acts to the axis about which the body may rotate. The force must be in a plane perpendicular to the axis of this rotation.

Rotational Motion

In order to discuss the amount a body has turned in its rotational motion, we must specify the angle between a line fixed in space and a line in the body. These two lines determine a plane. The hands of a clock are examples of rotating bodies. The hour hand at three o'clock has turned through an angle of 90 degrees since 12 o'clock. When this angle has increased through 360 degrees, we say that the hand has made one revolution. We can specify the orientation of a golf club by giving the size of the angle between the club and the line indicating its original position at the start of the rotation.

Angular and Rotational Acceleration

When a body is rotating about an axis, we say that the body has an angular velocity about this axis. This angular velocity is measured by the rate at which its orientation angle changes with time. Angular velocity is thus analogous to linear velocity. The second hand on a clock has an angular velocity of one revolution per minute, or 360 degrees per minute. Earth has an angular velocity of 360 degrees per day.

When the angular velocity of a body is changing, we say that it has an angular acceleration. This angular acceleration is measured by the rate at which the angular velocity of a body is changing with time. The angular acceleration is thus analogous to linear acceleration. The hands of a clock have zero angular acceleration, since they are rotating at an unchanging angular velocity. While a top is being started in its spinning motion, it has a positive angular acceleration, and while a top is gradually slowing down, it has a negative angular acceleration.

In rotational motion, the angular acceleration of a body is related to the torque acting on it. If there is no torque, there is no angular acceleration; the angular velocity remains the same. In linear motion, the acceleration of a body is proportional to the net force acting on it. And in an analogous fashion the rotational acceleration is proportional to the net torque acting on it. When a torque is applied to a body free to rotate but at rest, the body will start to rotate. As long as the torque continues to act, the body will spin faster and faster. In the start of the downswing of a club, the golfer's arms do not rotate at a constant rate like the hands of a clock, but they move at an ever increasing angular velocity. The angular acceleration depends on the torque applied to the golfer's arms. At the start of the downswing the arms have a positive angular acceleration, and later in the swing they have a negative angular acceleration; they slow down.

Moment of Inertia

The analogy between linear and rotational motion may be carried a step further. The linear acceleration of a body when acted upon by a constant force depends on its mass, which is a quantity proportional to its weight. The larger the mass, the smaller will be the acceleration for a given force. Similarly, we find that when a constant torque acts on a body, its angular acceleration will depend on the mass of the body and on how the mass is distributed in the body. The combination of mass and its distribution in the body is called its "moment of inertia." When the axis of rotation of the body is chosen such that the mass is distributed mostly near the axis, the moment of inertia will be relatively small. When the axis is chosen such that more of the mass is far from the axis, the moment of inertia will be larger. Thus the moment of inertia will depend on the choice of the axis.

This concept of moment of inertia can be demonstrated very easily with a golf club. When we hold a club at the grip end between two fingers and let the club hang so that the shaft is along a vertical line, we find that we may rotate it very easily about a vertical axis. But when we hold the club near the center of the shaft, where it balances between the same two fingers, and rotate the club about a vertical axis perpendicular to the shaft of the club, we find that the same torque produces a much smaller angular acceleration. The moments of inertia in the two cases differ by about a factor of ten. This same effect may be observed when we waggle a club first about the grip in the usual way and then waggle it holding the head of the club in our hands. Those who are interested in the definition of moment of inertia are referred to the Technical Appendix, Section 1. All that is needed here is a general feeling of how the distribution of mass of a body influences its rotational properties.

Angular Momentum

In discussing linear motion, we found it convenient to use the concept of momentum, the product of the mass of a body and its velocity. The analogous concept in rotational motion is the "angular momentum"; it is defined as the product of the moment of inertia of the body and its angular velocity.

The statement about rotational motion analogous to Newton's second law of linear motion is that the rate of change of angular momentum is proportional to the torque acting on the body. We have no need here for anything more than a general understanding of this statement. It means that if you swing a club with a large torque, its rate of rotation will increase rapidly, while if you swing it with a small torque, its rate of rotation will increase slowly.

Conservation of Angular Momentum

We have considered the principle of the conservation of linear momentum. The slow speed of recoil of a gun was explained in terms of this principle. The analogous conservation principle for rotational motion states that when no external torque acts on a body, its angular momentum is constant. This principle has illustrations in several common activities.

An ice skater can spin at a high rate if he starts at a slow rate on one leg with the other leg and his arms extended from the axis of the spin and then brings his legs together and his arms in close to his body. His angular momentum does not change in this maneuver, but his rate of turning, his angular velocity, increases in direct proportion to the decrease in his moment of inertia. Anyone may experiment with this effect on a smooth floor, but he should proceed with care. The conservation of angular momentum is used by high divers, ballet dancers, ski jumpers, and by cats, who use it when dropped upside down to land on their feet.

Centrifugal Torque

Again consider a golfer riding on a merry-go-round. Let him somehow be prevented from being thrown off by the centrifugal force acting on him in the rotating system. Let him extend his arms and a golf club out along a radius from the axis of rotation. He feels no torque on the club because the lever arm of the centrifugal force on the head of the club is of zero length. Next let him hold the club so that his wrists are cocked at ninety degrees. There is still a centrifugal force in the rotating system on the clubhead out along a radius from the axis. With this wrist cock, the centrifugal force will have a lever arm the length of the club. The golfer riding in the rotating system will feel this torque, and without a matching torque on the club by the golfer, the club will start to move out to a greater radius. The torque on the club is thus a centrifugal torque. In the swing of a golf club, this is the torque that brings the clubhead out to hit the ball. This torque becomes very large.

Work and Energy

A dynamic event such as the swing of a golf club may be analyzed in terms of forces and accelerations or in terms of torques and angular accelerations. A dynamic event can also be analyzed in terms of work and energy. The use of the concepts of work and energy allow us to understand a dynamic event in a different way than when we use the other concepts. We need therefore to develop these concepts, again on a qualitative basis, to allow us to look at the swing of a club from another point of view.

We shall begin by defining "work." Work is a word used in everyday language, but we shall use the word in a very particular manner. When we push on a body and the body moves in the direction we are pushing, we say that we are doing work on the body. The amount of work we do is given by the product of the amount of the force we apply to the body and the distance the body is moved in the direction of the force. When we lift a body we do work. When we stretch a spring we do work. When we throw a ball we do work. This concept can be extended to include work done when a torque turns a body through an angle. A child does work on a top when he gives it a spinning motion.

Potential and Kinetic Energy

When we have done some work, such as lifting a body or stretching a spring, we almost of necessity ask ourselves where the work went. We apparently feel the need to give the work an independent existence. We satisfy this need by saying that the work is stored as energy in the body we have lifted or in the spring we have stretched. We call this energy "potential energy." We do this because if we lower the body we have lifted or let the spring return to its unstretched condition, we have work done for us. We get back the work we put in earlier.

When we throw a ball we do work on the ball. The force we exert on the ball gives it an acceleration, and the ball is set in motion. When the ball is in flight, where has the work gone? We say that the work we did in setting the ball in motion is now stored as kinetic energy in the ball. When someone catches the ball, work is done by the ball, and the kinetic energy is removed from the ball. Energy can be considered in this context as stored work. The amount of stored work that can be usefully recovered in any situation depends on the circumstances and our ingenuity, If no work is done on a body or by a body, we say that the energy is conserved.

In the swing of a golf club, the intention is to do work by the large muscles of the body and in a long drive get as much of this energy as possible transferred to the golf ball. This transfer must take place according to the laws of physics.

Vectors

The preceding discussion is concerned with quantities that have magnitude and direction. Displacement, velocity, acceleration, momentum, and force have this character. Quantities of this kind are classed as "vectors." Vectors of a given kind may be added and subtracted according to a certain rule sometimes called the triangle or parallelogram rule. An example will make this rule clear.

Consider two displacements, one at right angles to the other. Something might be moved north one foot, and then it might be moved east one foot. A single displacement equivalent to the other two would be one in a northeasterly direction from the starting point of the first displacement to the end point of the second displacement. This latter displacement is the vector sum of the first two displacements. The original displacements do not have to be at right angles to one another. The reason the rule of addition of two vectors is called the triangle or parallelogram rule is evident.

As the result of this rule for adding vectors we may reverse the addition process and take any vector and resolve it into what are known as its components.

Components are vectors that when added give the original vector. Components are usually taken at right angles to one another. One component might be along the horizontal direction and the other along the vertical direction. Thus we may speak of the horizontal component of a vector and the vertical component of a vector.

If a vector represents the velocity of a golf ball in flight, then the horizontal component would tell us how fast the ball is moving in the horizontal direction, and the vertical component would tell us how fast the ball is moving in the vertical direction. These components are found simply by taking the right-angle projection of the vectors along these particular directions. The component of a vector in a particular direction is sometimes called the effective part of the vector in that direction.

We shall find that components are useful in discussions concerning vector quantities. As an example, in the work done by a child pulling a sled horizontally along through the snow, the displacement of the sled is horizontal, and the displacement vector has no vertical component. However, the rope between the child's hand and the sled is at an angle to the horizontal. The force in the rope, its tension, has both a horizontal component and a vertical component. The horizontal component of the force on the sled is in the direction of the displacement, and according to the definition of work, must be multiplied by the displacement to obtain the work done by the child. The vertical component of the force does not enter into the determination of this work.

Summary

The intention of the discussion of this section is not to give the reader a working knowledge of dynamics, but rather a speaking acquaintance with the subject. He should be acquainted with Newton's three laws of motion. He should recognize the relation of force to acceleration and the relation of torque to angular acceleration. He should have a general idea of centrifugal force, work and energy, conservation of momentum and angular momentum, and moment of inertia, and should recognize the third-law nature of force and torque. He should realize that the general ideas and concepts that have been introduced may be used to write quantitative mathematical expressions for any dynamics problem.

INDEX

A

Acceleration, 16, 49
Acceleration, angular, 4, 10, 27, 49, 173, 180
Acceleration, centripetal, 111
Acceleration, gravitational, 150, 152
Acceleration, linear, 7, 173
Aerodynamic forces, 65, 68, 86
Airflow patterns, 70
Angular velocity, 9
Armour, Tommy, 27

B

Backswing, 57
Backswing, reduced, 35
Ball slides and rolls on clubface, 80
Ball, inelastic, 79
Ball, position of, 55
Beard, Frank, 27
Bernoulli, Daniel, 67, 68
Boros, Julius, 28
Boundary layer, 69

C

Clubhead meets the ball, 139
Clubhead speed, 15
Clubhead speed, effect of gravity, 34
Clubhead speed, effect of shift, 34
Clubheads, perimeter weighting, 2, 90
Clubs, perfect matching, 99, 101
Clubs, relaxed matching, 103
Clubs, swing-weight matching, 98
Coins, tossing of, 119
Collision theory, 78, 85, 176

Competing effects of lift and drag, 73
Computer, v

D

D-Plane, 83, 86
Data, experimental, 5, 151, 152
Developing your own golf swing, 51, 64
Differential equations, 150, 156
Dimples increase turbulence, 71
Discus, 47
Downswing, 58
Drag, 71, 82

E

Edgerton, Harold E., 5
Efficiency of the golf swing, 44
Elasticity, 79
Energy, 20, 42, 161
Energy, calculation of, 42
Energy, conservation of, 42
Energy, heat, 19
Energy, kinetic, 19, 44, 152, 183
Energy, potential, 42, 43, 152, 183
Euclid, 84
Experiments, motion of rod, 21
Expert Golfers, remarks by, 27

F

Factors affecting drag, lift, and spin, 83
Feel of swing, 64
Fermat, Pierre de, 118
Fisher, Jerry, ix
Flex on loft of club, 116